Post-War Eugenics, Reproductive Choices and Population Policies in Greece, 1950s–1980s

Alexandra Barmpouti

Post-War Eugenics, Reproductive Choices and Population Policies in Greece, 1950s–1980s

palgrave
macmillan

Alexandra Barmpouti
Athens, Greece

ISBN 978-3-030-03567-9   ISBN 978-3-030-03568-6   (eBook)
https://doi.org/10.1007/978-3-030-03568-6

Library of Congress Control Number: 2018968349

Cover credit: Overcrowded Downtown Building Abandon Apartment © Rawpixel Ltd/Alamy Stock Photo; Side View Of Silhouette Of Pregnant Woman Against White Background © Supachok Pichetkul/EyeEm/Getty Images
Cover design by Fatima Jamadar

This Palgrave Macmillan imprint is published by the registered company Springer Nature Switzerland AG
The registered company address is: Gewerbestrasse 11, 6330 Cham, Switzerland

PREFACE AND ACKNOWLEDGEMENTS

The purpose of this book is to identify and fill the gaps in the historiography of post-war Greek eugenics. As will be explained in the introductory Chapter 1, historians were many times reluctant to undertake such an attempt because on the one hand the name of eugenics became a taboo after the revelation of Nazi crimes and on the other hand because the bibliographical sources for the history of eugenics during the first half of the twentieth century are much richer.

After completing my master thesis on the ethical considerations of eugenics concerning modern reproductive technologies, I identified a gap between the history of eugenics during the first half of the twentieth century and the modern bioethics debates. It was very difficult to find bibliographical sources concerning the history of eugenics between the 1950s, when preoccupation with Nazi eugenics came to an end, and the 1980s when arguments against the eugenic use of biotechnology came to light in the context of bioethics. I was then motivated to do research in this time frame.

I selected the geographical region of Greece because bibliography on Greek eugenics at any period was scarce and because of the rare example of the establishment of the Hellenic Eugenics Society in 1953.

I was fortunate to be the first to explore the Nikolaos Louros Papers and Archive with the valuable help of Constantinos Trompoukis, Associate Professor of History of Medicine at the Department of History of Medicine at the University of Crete and President of N. Louros Foundation and Mrs. Anna Manidaki, Scientific Assistant at the same department. When I started digging the treasure of this archive, I realised that its information could not be found in published bibliography. The most important element of knowledge was the untold history of the collaboration among the Greek, British and American eugenicists which is analysed in this book. Louros archive guided me to more archives; the most crucial of all was that of Clarence Gamble. The outstanding result of the comparative study of these archives was the fact that I could see the whole picture of their correspondence because both sides kept the letters they send to each other. To a great extent, the study of the correspondence enlivens history and permits the reader to make a journey in history and most importantly to evaluate the facts from both sides. An additional advantage of researching Gamble's archive was the fact that he and his delegates in Greece wrote detailed reports on the situation in the country regarding eugenics, contraception, sex education, the function of the Hellenic Eugenics Society, the activities of women's clubs and other important, unpublished information.

In this book, the archival material was combined with publications of the period under consideration but also with more recent bibliography on the history of eugenics, birth control, demography and modern history.

In terms of bibliographical sources, the book is divided into three main parts. The introductory Chapter 1 and Chapter 2 were mainly produced by the study of published books and articles but were also based on legal texts. Chapters 3 and 4 almost exclusively relied on archival material. Finally, Chapter 5 was constructed upon published books and articles which refer to the Hellenic Eugenics Society's activities and publications in combination with legal texts. Thus, the book offers a wide spectrum of sources in order to provide the reader with a holistic view of post-war eugenics.

As the title suggests, the book contains information and analysis of post-war eugenics, issues of reproduction and population policies. I tried to understand the mentality and attitudes of the Greek and foreign eugenicists and the Greek women. The goal of this book is to demonstrate the continuity of eugenics throughout the twentieth century. Either in the USA, in Greece or in Japan, eugenicists tried to manipulate the reproductive choices and manage the quality and quantity of population.

I am grateful to my former supervisor of my Ph.D. thesis at Oxford Brookes University and beloved friend Marius Turda, whose comments and moral support were invaluable. Moreover, I owe special thanks to my family and friends for their encouragement during the writing process of this book.

Athens, Greece   Alexandra Barmpouti

CONTENTS

1   Introduction   1

2   Health and Hygiene, 1900–1950   21

3   The Hellenic Eugenics Society   45

4   The International Network   85

5   Eugenic Concerns   133

6   Conclusions   191

Index   203

ABBREVIATIONS

AES	American Eugenics Society
AFGS	Association of Female Greek Scientists (Συνδεσμος Ελληνίδων Επιστημόνων)
AMA	Athens Medical Association (Ιατρικός Σύλλογος Αθηνών)
BES	British Eugenics Society
HES	Hellenic Eugenics Society (Ελληνική Εταιρεία Ευγονικής)
ICPP	International Committee on Planned Parenthood
IPPF	International Planned Parenthood Federation
NCGW	National Council of Greek Women (Εθνικό Συμβούλιο Ελληνίδων)
NUSE	National Union of Sanitary Education (Εθνικός Σύνδεσμος Υγιεινολογικής Διαπαιδαγωγήσεως)
PIKPA	Patriotic Institution of Social Welfare and Awareness (Πατριωτικό Ίδρυμα Κοινωνικής Πρόνοιας και Αντίληψης - ΠΙΚΠΑ)
UN	United Nations
WHO	World Health Organisation

CHAPTER 1

Introduction

Eugenics is often associated with Adolf Hitler's dream of building a pure, master race. Racial hygiene ideology (*Rassenhygiene*), which had been developed in Germany since the early twentieth century and culminated during the Nazi era, was grounded in racial classification. The Aryan race to which the Germans belonged stood on the top of the scale. Thus, their goal was to purify and strengthen their race by multiplying its members while eliminating the presence of foreign races in the German nation. Indeed, during the Nazi period, the forced reproduction of 'superior' Aryans and the extermination of 'inferiors' became compulsory state policies resulting in hundreds of deaths of Jews, Gipsies, disabled, or any other group of people alien or harmful to their racial growth and purification (Weindling 1989; Kühl 1994; Proctor 1998; Pichot 2000; Kallis 2007).

Racism was also supported under other classification scales in Europe and the USA during the same period, mostly expressed as biological racism. The fundamental element of biological racism was the belief that race was initially constructed and should further be developed upon the common biological past of a certain group of people. Therefore, race was not always identified with a specific nation but its notion extended to larger groups, such as the Nordic race, and at times to the entire human race.[1]

Without denying the obvious relation between racism and eugenics, it would not be accurate to identify them. Eugenics indeed embraces

© The Author(s) 2019
A. Barmpouti, *Post-War Eugenics, Reproductive
Choices and Population Policies in Greece, 1950s–1980s,*
https://doi.org/10.1007/978-3-030-03568-6_1

the distinction between superior and inferior groups of people or individuals but is not limited to racial biological classification and segregation. For instance, eugenics often dictated the segregation of people due to socially inappropriate behaviour or poverty to name a few non-biological criteria. In the same context, eugenics supported the proliferation of 'worthy' individuals based on their upper social class and not strictly on their valuable genetic inheritance. Undoubtedly though, many eugenicists were racists and many racists were eugenicists, particularly in the beginning of the twentieth century. As Turda argued, '...the fundamental reality was that eugenics was born into a period when European and American societies thought in terms of racial categories, and believed in the existence of 'superior' and 'inferior' races' (Turda 2010c, p. 66). Added to this, the fact that the Nazi practices were so radical both in eugenic and racist terms is the reason why the association among Nazism, racism and eugenics became the most popular example of eugenics in people's minds.

However, eugenics existed before and after Nazism, related to many different ideologies and localities making its definition a difficult task. Yet, every form of eugenics has been rooted in its core ideology, the control of reproduction. The spectrum of eugenics ranges from individual reproductive choices to state intervention in mating and reproduction. Decades earlier than the Nazis, Sir Francis Galton developed the theory of eugenics to become a panacea for the social injuries of Britain. Eugenics was linked to the works of other British scholars, such as Thomas Robert Malthus' population theory, Jeremy Bentham's utilitarianism and, of course, Charles Darwin's theory of evolution. Galton incorporated statistics in genetics and did extensive research on family trees of famous and important British families to prove that human qualities, such as intelligence, were inherited (Galton 1869). As a supporter of biological determinism, Galton argued that each state was obliged to protect its superior, 'biologically valuable' citizens by providing them with the available means to reproduce more and more descendants, simultaneously refraining the 'biologically lower classes' from reproduction. The aim of the implementation of eugenics was to create a society with biologically robust and intellectually superior people who would gradually become even better because they were anticipated to genetically mix and socialise with superior fellow-citizens, thus ameliorating the human species as a whole. Eugenics was immediately embraced by the Americans who went further by implementing eugenics through state policies. The survival of the fittest became a state concern and relevant legislation was passed in many states (Stern 2005). Either as an academic discussion or a state policy, eugenics flourished during the first half of the century and gradually gained ground beyond the Anglo-Saxon world. Soon, it became an international phenomenon reaching cultures as variant as Asian and Latin (Adams 1990; Stepan 1991; Turda and Weindling 2007; Bashford and Levine 2010; Turda and Gillette 2014). The broad participation of scholars in the human betterment debate was evident in various cultural and political contexts around the world thus making the first half of the twentieth century the golden era of eugenics. During the same period, eugenicists formed associations, organised international conferences and made eugenics a scientific discipline taught in European and American universities.

However, the abuse of eugenics ideology coupled with racism and the imposing authority of a dictatorship led to the historical phenomenon of the Nazis. Following the public outcry provoked by the disclosure of Nazi eugenics, public discussions about social and racial purification came to an abrupt end after the end of the Second World War. Scientists and politicians avoided any connection with the tarnished ideology of eugenics and tried to conceal their pre-war activity. In this context, universal conventions were signed in order to prevent the repetition of similar practices in the future. These were the Convention on the Prevention and Punishment of the Crime of Genocide (UN 1948a) and the Universal Declaration of Human Rights (UN 1948b). As eugenics was at the time associated with Nazism and racism, these universally agreed conventions condemned it, albeit not explicitly. The international eugenics movement did not cease to exist; what essentially changed was that the word 'eugenics' was limited to private discussions. The word 'eugenics' was generally avoided and replaced by the new term 'human genetics'. In Britain, the birthplace of the modern eugenics movement, the journal *Annals of Eugenics* was renamed as the *Annals of Human Genetics* in 1954, but *The Eugenics Review* became the *Journal of Biosocial Science* much later, in 1968.

Nevertheless, eugenics remained a largely supported ideology well after the 1950s. Being a versatile and adaptive ideological structure, eugenics was integrated in socio-economic issues, such as the confrontation of overpopulation by birth control. Without losing its grip to the control of fertility and human reproduction, eugenics embraced the international birth control movement. Facing its degradation, post-war eugenics changed focus from racism to family planning strategies (Glass 1966; Connelly 2008; Bashford 2014). Eugenicists supported birth control to allegedly protect some countries from overpopulation and the Earth from its catastrophic consequences (Ramsden 2002, 2009). However, the goal remained the same, namely the manipulation of human evolution through the control of reproduction. The issue of overpopulation was promoted by a massive publication of related literature terrifying the public with the threat of a dystopian future on the overpopulated Earth. Bestsellers of eugenicists, such as Fairfield Osborn's *Our Plundered Planet* (Osborn 1948, 1962) and Paul Ehrlich's *The Population Bomb* (1968) played a significant role in propagating eugenics veiled under overpopulation concerns (Vogt 1948; Descochers and Hoffbauer 2009). Furthermore, in the 1970s, there was a tendency in film production to deal with similar issues, such as the popular film *Zero Population Growth* (1972).

In addition, many international alliances emerged to tackle demographical and ecological issues, such as the Population Council in the USA, the International Planned Parenthood Federation (IPPF) and United Nations Educational, Scientific and Cultural Organization (UNESCO), mainly supported by eugenicists such as Julian Huxley, Margaret Sanger, William Vogt, Carlos Blacker and many others (Stern 2005; Bashford 2014). The international birth control movement was initiated mostly by a group of Americans and British eugenicists (Stern 2005). It was not a coincidence that, contrary to the global trend of eugenics condemnation, the British Eugenics Society remained active until 1968 and the American Eugenics Society continued its activities until 1972.

Although a small number of eugenics societies continued their activities after the Second World War, the Hellenic Eugenics Society (HES) was the only one which commenced its activities during the post-war period. However, nowadays, the HES is still unknown to the general public and scholars, both in Greece and abroad. Why was this society founded then and not earlier, as was the case in other European countries? Moreover, why establish a Eugenics Society at a time when most scientific societies were gradually distancing themselves from eugenics? These considerations notwithstanding, there are a number of reasons why the HES was established at the beginning of the 1950s in Athens, as will be discussed throughout the book, together with a number of other topics related to eugenics, demography and medical genetics.

The study of the Greek example is important to the history of eugenics because it represented one of the few cases where eugenicists overtly expressed their views after the Second World War. The HES organised a series of successful public conferences from the 1950s to the 1980s, discussing the major issues of the time, such as overpopulation, family planning, genetic diagnosis and counselling. Contrary to the Greek state's pro-natalist policies and despite the decreasing birth rate, Greek eugenicists put effort to disseminate family planning methods. The control of the quality of reproduction and the quantity of the population was the ultimate expression of eugenics during the post-war period.

Above all, the case of Greek post-war eugenics is valuable to the study of the history of eugenics because it proves the continuity of eugenics during the post-war period which is less discussed by historians. This book builds on existing literature on interwar eugenics and post-war birth control by eventually providing a linkage between them, thus expanding our knowledge of reproductive choices and population policies during the post-war period. Furthermore, the investigation of the collaboration among Greek, British and American eugenicists reveals that the HES was not limited in the national borders of Greece but was part of the transnational eugenics movement veiled under the dissemination of birth control policies and family planning strategies. In particular, this book provides unprecedented reasoning of the continuity of eugenics during the post-war period because its argumentation is based on original archival material. The importance of the disclosure of this material is twofold because not only reveals the content of the archives but also provides a research study derived from the combination of Greek, American and British archives.

Consequently, this book shows that the HES was not a rare and outdated exception among similar eugenics societies in Western Europe and the USA, which flourished at the beginning of the twentieth century. As will be discussed, it followed the international tendency to popularise birth control in conjunction with demographical concerns. Taking into account, the rich historiography of pre-war eugenics, it is acknowledged that the HES distanced itself from ideas and practices which tarnished eugenics ideology, such as ideas of racial purity, sterilisation and especially 'euthanasia' while focusing on issues of family, reproduction and demography.

Social and biological degeneration was mostly attributed to irresponsible childbearing and to the lack of preventive medicine. Contrary to the intervening role of eugenicists during the first half of the twentieth century, post-war eugenicists were mostly interested in guiding individual reproductive choices and cultivating the 'procreation instinct' and parental responsibility (Ziegler 2008; Kline 2001). Therefore, post-war Greek eugenicists also focused on offering family planning advice before and during marriage and guidelines for the proper raising of children, irrespective of the social status of individuals. Similarly to other countries, emphasis was placed on the protection of pregnant women and newborn children, improvement of living conditions, preventive medicine and public hygiene, individual marital and reproductive choices and control of the (female) body (Gillette 2007).

The main difference between the development of eugenics during the first half of the twentieth century and its institutionalisation during the second half was the focus on family planning. In the former case, eugenics was expressed through the collective purpose of racial regeneration and protection (Trubeta 2010). Racial superiority, as expressed during the interwar period, was no longer part of the post-war eugenics rhetoric. Additionally, the sociopolitical conditions had changed. The first half of the century included the redistribution of territories and the creation of new borders for many European countries, including Greece. In the context of nation-building, eugenicists employed race as a fundamental common trait to construct the new identity of their nation (Turda 2010a).

On the contrary, during the post-war period, endorsing racism and nationalism was not an effective means to promote eugenics. In the 1950s, the Greeks were mostly interested in personal health and social restoration which became imminent after many years of continuous warfare. Added to this, the composition of eugenic associations had changed. A shift from physical anthropologists to physicians, in particular gynaecologists, was observed. Instead of praising the glorious past and trying to protect racial purity, post-war eugenicists opted to invest in the future of the country. They made serious attempts to reform and modernise state policies and society. To this end, they aimed at strengthening the health of the Greek citizens and promoting social progress; while also encouraging healthy births. Therefore, priority was given to the individual reproduction control. Social prosperity and biological regeneration were to be achieved by conscious reproduction.

Post-war Greek eugenics was linked to the modernisation efforts of the Greek society and the foreign influence in this process. For instance, it was the American funding that aided the improvement of the health sector, including the building and renovation of major hospitals in Athens, such as the 'Alexandra' Maternity Hospital. At the same time, state authorities and health professionals promoted preventive public health and hygiene policies. Simultaneously, the state adopted laws for the transformation of hygienic services and a law for public education on health and hygiene; and physicians and health professionals tried to disseminate their knowledge to the public—the general aim was the regeneration of society. Law 2032 was a representative example of health and hygiene promotion. It was adopted in 1952 and announced the formation of a new public service which would provide public education on preventive health and hygiene (Official Government Gazette 1952). This new service was meant to facilitate the establishment and function of institutions, associations or individuals that accorded with its purpose.

Above all, physicians, particularly gynaecologists and paediatricians, played an important role in the rise of eugenics and birth control movement in Greece during this period. Influential personalities, such as Nikolaos Louros, Maro Kanavarioti, Vasilios Valaoras, Spyros Doxiadis, Georgios Pantazis and Panayiotis Panayiotou, were involved in the establishment of the HES and in shaping its activities, both in Greece and abroad. Louros is considered as the 'heart' of the HES but the participation of eminent physicians in the activities of the HES is equally important. Most of them worked in public health institutions and were associated with the big hospitals of Athens. The general tendency was to put the authority to strengthen the Greek society, physically and spiritually to the hands of the much-respected health professionals. As was often observed, physicians' national protectionism stemmed from their alleged ability and obligation to promote social prosperity and robustness through eugenics education (Turda 2010b).

Furthermore, the changes in demographic patterns and the ongoing emancipation of women were also important factors in the development of eugenics and birth control (Barmpouti 2015a). As was clearly depicted in the demographic and statistical analyses of the time, contrary to the global overpopulation problem, Greece experienced demographic stability during the 1950s. This was primarily due to the loss of human capital during the Second World War and the Greek Civil War, high mortality rates and the increase of induced abortions. With the exception of an increase in births between the 1960s and the 1970s, the birth rate followed a downward trajectory during the following decades (Burnova 2016). As an overall image, the birth rate from the 1950s to the present-day dropped from 28.8% in 1951 to 14.4% in 2010. Although mass emigration and/or high rates of mortality played a significant role to a population, the decline of births was often the most important of all, because it often led to population stability and might inhibited the population from renewal in future generations (Soloway 1995). When this demographic problem became apparent in Greece, political authorities and some members of the scientific community opposed any birth limitation practice, simultaneously adopting pro-natalist policies. There was, however, another group in the medical and scientific community who were influenced by the global problem of overpopulation—which was also becoming central to demographic research during this period—and who embraced neo-Malthusianism and warned against overpopulation, food scarcity, unemployment and space limitation. As a result, there were those who opposed contraception, perceived as a birth limitation method, and those who supported it, translating it into conscious family planning and pregnancy-spacing.

The polarisation of opinions about family planning was intensified by the issue of induced abortions, which became one of the most important socio-medical and demographic problems in Greece after 1950 (Valaoras et al. 1969; Valaoras 1969). Induced abortion, the use of contraceptives and birth decline formed a vicious circle (David 1992). Some argued that there was no infertility issue in Greece, but that induced abortions led to birth decline; while others claimed that the use of contraceptives led to birth decline because women used them to avoid conception. At the same time, the proponents of birth control argued that a woman should control her reproduction, having the desired number of children at the desired time. They argued, paradoxically, that the only way to avoid abortion was contraception, and that contraceptives did not lead to a decrease in births, but only to better planning of their family. Greek gynaecologists were also divided into two groups: those who opposed abortion and promoted contraceptive techniques and those who indirectly supported induced abortion because they earned large sums of money from performing it. Finally, the absence of sex education in Greece should not be overlooked in the discussion of the socio-medical problem of abortions and unwanted pregnancies. The lack of family planning advice and sex education narrowed the reproductive choices of Greek women, often choosing abortion as the only means to deal with an unwanted pregnancy (Barmpouti 2015b).

The legal framework of the country for abortion and contraception was constructed upon the pro-natalist perspective, which condemned both induced abortion and female contraception. In fact, this attitude indirectly imposed state biopolitics on the reproductive freedom of the individual. Albeit illegal, abortion was equally performed by married and single women, both in urban and in rural areas. As was argued at the time, induced abortion was a condition of 'legal illegality', because actually there were no prosecutions for the violation of the existing law which prohibited abortion (Roukas 1979; Halkias 2004).

The legalisation of abortion, the use of female contraceptives and family planning advice occurred during the 1980s. Due to the fact that family planning advice was illegal until then, it is observed a lack of literature about it before that time. However, there is abundant scholarship on the history of family planning in Greece after the 1980s (Ioannidi-Kapolou 2004; Paxson 2002, 2004; Tseperi and Mestheneos 1994). This book is the first attempt to portray the history of family planning advice in Greece during the period from the 1950s to the 1980s. Although family planning advice was officially and legally provided by gynaecologists from 1980, in reality, it was during the 1950s that eugenics and family planning began to gain wide support among physicians and academics in Greece. The change of the family law in the 1980s was the hallmark of a new era in reproduction politics, raising a wide range of new issues, both medical and ethical.

Although the history of family planning techniques dates back to the 1950s, it gained publicity much later. In the 1980s, many countries, mostly in Europe and the USA, passed relevant laws, such as the right to contraception and abortion. But what happened in the meantime? What was the role of eugenicists during the period from the 1950s to the 1980s? This book intends on shedding light on the history of eugenics during this obscure period when eugenics had already acquired a negative connotation. The period under discussion coincides with the entire period of the existence of the HES, from its establishment in 1953 to its gradual demise in the early 1980s. In addition, this particular time period includes very important breakthroughs related to eugenics, such as the advance of human genetics, the rise of the feminist movement and technological innovation. The book will cover for the first time the history of contraception in Greece during the 1950s and the 1960s, when the use of female contraceptives was forbidden. It will show the course which women and gynaecologists followed until the passing of the law which permitted family planning advice and the use of contraceptive techniques. The fact that it was a group of eugenicists who struggled and finally succeeded to pass this law adds value to the argument that birth control was ideologically based on eugenics. Besides, the leaders of the international birth control movement were renowned eugenicists. In many ways, this book proves that eugenics did not disappear after the Second World War.

As already mentioned, the condemnation of eugenics in the 1950s was the threshold between its thriving first period and the post-war period of veiling under birth control. Since the early twentieth century until 1950, one encounters abundant scholarship on eugenics ideology, the passing of eugenic laws, the establishment of eugenics societies and institutions and a flourishing academic interest equally expressed by natural and social scientists and others. Many scholars divide the history of eugenics into 'old' and 'new', often referring to the difference between the collective purpose of social improvement of the first period and the individualistic purposes of genetic improvement during the post-modern era (Allen 2001; Epstein 2003; Duster 2003). However, as Ekberg argued, the ideological foundation remains the same (Ekberg 2007).

Not surprisingly, eugenics scholarship mostly focuses on the historical period from its public inauguration in the late nineteenth century to the 1940s. For the reasons stated earlier, the historiography of Nazi eugenics has been particularly rich (Weindling 1989; Kühl 1994; Proctor 1998; Pine 1997; Pichot 2000; Weindling 2004). On the contrary, post-war scholarship on eugenics is inadequate because scholars of the time incorporated their eugenic ideas into debates about overpopulation and other relevant demographic issues (Dowbiggin 2002). In addition, most historians rarely discuss post-war eugenics, either because they claim that it ceased to exert any influence on society or because it is easier to do research on interwar eugenics when it flourished and was endorsed officially by many scholars and politicians (Bashford and Levine 2010). Even so, historians admit that eugenics did not cease from existence (Bashford and Levine 2010; Cogdell 2004; Ekberg 2013). Furthermore, it is identified a gap in eugenics literature between the 1940s and the 1990s when eugenics debates emanated from the advance of reproduction genetics and human enhancement technologies in the context of bioethics. Duster's classic *Backdoor to Eugenics* was one among many examples (Duster 2003). The lack in post-war historiography of eugenics during the period under discussion of this book does not imply that eugenics had disappeared but indicates that there is room for more research.

Similarly to the history of eugenics generally, the Greek secondary literature refers almost exclusively to the first half of the twentieth century. As was the case in other countries, it was during the first half of the twentieth century that the Greek eugenicists expressed themselves in academia, publications and public debates suggesting eugenics as the most effective tool of racial improvement. In this context, eugenics was identified in debates about health and hygiene restoration and improvement of the society, mostly endorsed by physicians (Karakatsani and Theodorou 2010, 2012). Moreover, there was an active participation of scholars, mostly anthropologists, in discussions promoting the eugenic control of the nation. The Greek Anthropological Association, in particular, an esteemed association of scholars provided the space for eugenic discussions and hosted debates on racial purification (Trubeta 2007, 2013).

A probable explanation for the scarcity of post-war eugenics historiography, at least in Greece, could be explained by the problematic legacy of eugenics globally. However, in this case the most important step towards the study and dissemination of eugenics was the foundation of the HES in 1953. So far, there is only one publication referring to the HES but in the context of the eugenic concept of race in Greece (Trubeta 2013). The author devoted a few pages on the activities of the HES based on the published material, such as the published minutes of the HES's public conferences. Among others, the author correctly pointed out that the HES formed a part of the overall modernisation effort of the Greeks after the wars. The participants of the public conferences were renowned professionals and the most prestigious figure was that of Nikolaos Louros, the President of the HES for twenty years. The book examines eugenics mostly with regard to racial anthropological debates.

This book, however, goes beyond this study. That is because information and analysis of this volume derive not only from published material but also from unpublished archival material. Louros possessed a valuable archive of the activities of the HES, ranging from the original statutes of the society and its internal structure to personal correspondence with Greek and foreign eugenicists. Inter alia, the exploration of Louros' archive reveals for the first time the motivation behind the resurgence of eugenics in Greece during the post-war period. Research showed that foreign encouragement was decisive for both the establishment of the HES and the development of eugenics and family planning in Greece during the 1950s. In particular, the IPPF was one of the foreign institutions working closely with the HES; its regional department dealing with Europe, the Near East and Africa was established in London in 1952. Furthermore, the American demographer Pascal K. Whelpton visited Greece in 1952 and evidence shows that he was the one who motivated the Greek physicians and demographers to form a Eugenics Society. Equally important, Louros' archive includes the information that the first meetings of physicians who were interested in establishing a eugenics society were held in the premises of the Athens Medical Association and the first President of the HES was not Louros, as is widely known, but the then President of the Athens Medical Association, Athanasios Mantellos. Such information was impossible to be found elsewhere than Louros' unexplored and unpublished personal archive.

On top of that, another archive concerning eugenics and birth control in Greece comes to light for the first time in this book. Clarence J. Gamble, a prominent American birth control advocate and eugenicist, became interested in offering assistance to those involved in the dissemination of birth control in Greece in the 1950s. The fact that Gamble's personal archive provided an external overview of the situation in Greece regarding birth control practices was an invaluable contribution to the current research. Gamble and his associates carried out research in Greece and developed contacts with Greek eugenicists. They filled reports on the country discussing their activities, a unique guide to the bigger picture of the history of post-war international and Greek eugenics at the same time. Thus, the originality of this book derives from the combination and analysis of historical archives which reflect on the post-war eugenics activity of international alliances, such as Gamble's associations and the IPPF, and local eugenics societies, such as the Greek and the British. Furthermore, the combination of Louros' and Gamble's archives provided us with complete correspondence from both sides. Tellingly, Louros' letters were found in Gamble's archive and Gamble's response in Louros' archive and vice versa.

Without Louros' and Gamble's historical archives, any discussion about the development of eugenics in Greece was mostly hypothetical. The HES's publications are few and secondary bibliography is scarce. As will be shown in the chapters that follow, the combination of Louros', Gamble's, Dorothy Brush's and the British Eugenics Society's archives makes this book a unique contribution to the history of eugenics on the international level. Tracing back to the personal correspondence of eugenicists, we encounter a global network, led by the British and the Americans, which orchestrates the international birth control propaganda.

Bridging the gap between pre-war eugenics and contemporary discussions about the moral implications of genetics, the objective of this book is to elucidate the less-discussed period between the 1950s and the 1980s. It discusses the history of eugenics in Greece during the post-war period and argues that eugenics remained an important part of debates on demography, family planning and social progress, more generally. The example of Greece was not exceptional, but, as the relationships of the HES with British and American Eugenics Societies and the IPPF reveal, was part of a global network. As such, this book sheds light on the neglected topic of the existence of post-war eugenic movements, both in its national and international contexts.

Furthermore, it proves the continual development of eugenics in Europe and the USA during a period when it supposedly went into disrepute and challenges the belief that the eugenic mentality in Europe and the USA was abandoned after the Second World War. Although there are still voices claiming that eugenics disappeared during the 1950s, this book demonstrates the contrary. It provides important information and analysis for both the Greek and international historiography of eugenics. Eugenics and family planning are explored in the light of the demographic problems Greece experienced during the post-war period and up to the 1980s. Therefore, we can safely argue for the uninterrupted continuation of eugenics deriving from historical evidence.

1   GREEK TERMINOLOGY

The Greek language permits flexibility in the choice of words describing certain eugenic terms. There are many words with the same meaning, each having a positive, negative or 'ethically' neutral sense. For example, there are the two words for 'abortion': the word έκτρωση, which has a negative sense, and the word άμβλωση (Bampiniotis 1998), which is milder and used in medical/academic terminology. Likewise, there are two words for 'eugenics'. Both Greek words, ευγονία and ευγονική, are translated as 'eugenics' into English. In fact, ευγονική refers to the science of eugenics, the branch of genetics that studies the ways for physical or spiritual enhancement of humankind by the application of the laws of genetics and heredity; whereas *ευγονία* means to have healthy and some-times many descendants; and to be fruitful or, indeed, to be productive (Liddell and Scott 1940; Bampiniotis 1998). The latter choice is closer to the Ancient Greek meaning, as used by Plato (380BC), which simul-taneously is a positively charged term (Plato 1948). The word *ευγονική* sounds more scientific, albeit having a negative bias. The use of appro-priate terminology was essential in introducing the new HES. Therefore, although its name in Greek was *Ελληνική Εταιρεία Ευγονικής* and the word *ευγονική* which sounds more formal was selected, in the first public lecture on behalf of the HES, Nikolaos Louros, the then President of the HES, tactfully talked about *ευγονία*, assuming that the general public would find it easier to identify the positive experience of eugenics with this term.

The word 'Hellenic' instead of 'Greek' was also chosen on purpose. Hellenic is associated with Ancient Greece, whereas Greek is the word that foreigners use to describe the Greek nation. *Έλληνας* is the Greek word for 'the Greek' which stems from the word Ελλάς (Hellas), not from the word Greece. Moreover, the word 'Hellenic' alludes to the entire Greek nation, both the inhabitants of Greece and the Greek dias-pora. Aiming at building a formal and academic profile for the Eugenics Society, the word 'Hellenic' might be used because again it is more for-mal than 'Greek'. Moreover, the word 'Hellenic' shows the national/racial continuity from antiquity to the present. The admiration for Ancient Greece, in conjunction with the desire of the HES to strengthen the Greek national identity and improve the health and prosperity of society to become a robust nation, might be the reasons why 'Hellenic' and not 'Greek' was used.

The Ancient Greek legacy was praised by eugenicists at large and not least by the 'father of eugenics', Sir Francis Galton. As he put it: 'The ablest race of whom history bears record is unquestionably the ancient Greek, partly because their master-pieces in the principal departments of intellectual activity are still unsurpassed, and in many respects une-qualled, and partly because the population that gave birth to the crea-tors of those master-pieces was very small' (Galton 1869). Based on the above considerations regarding the essential meaning of certain words describing eugenics in Greek, we can surmise that the HES had con-sciously chosen its name as an illustration of both the historical continu-ity and ideological specificity it arguably represented.

2   STRUCTURE

The second chapter introduces the reader to the historical and demographic context of Greece from the beginning of the twentieth century until 1950. The consecutive wars from 1897 to 1949 unavoidably led to many changes both in population and territory. Influenced by the popularity of eugenics in Europe and the USA while concerned with the health of the population, Greek scholars and scientists brought eugenics to light. The chapter addresses issues of eugenics, health policies and hygiene implemented in Greece during the first half of the century. More specifically, it refers to the relevant legislation which shaped the activities and organisation of the public health sector, including practical solutions for health and hygiene, such as the student health card, and the introduction of eugenic views by some leading physicians and pedagogues.

The third chapter discusses the establishment of the HES in Greece. During the first period, from the beginnings of 1953 until the end of 1954, the HES was attached to the activities of the Athens Medical Association (hereafter AMA). The preliminary meetings were held at the premises of the AMA and chaired by Athanasios Mantellos who was simultaneously President of both the AMA and the HES, holding both posts until August 1954 when he was replaced by Nikolaos Louros. The first period included a series of meetings with regard to the structure, aims and the scheduling of the HES's activities. In this context, the participants of those meetings decided the content of the statutes too. An important component of the eugenics movement in Greece, the text of the official statutes is also examined as the quintessence of post-war Greek eugenics mentality. The second and most fruitful period of the HES began when Louros became its president, and the HES was transferred to the Alexandra Maternity Hospital in 1954, only a year after its establishment. Consequently, the HES separated from the AMA and became an independent association. Louros made endless efforts to establish the HES as a respected institution. He exploited his important connections in academia, politics and health services to fulfil his aims. The hallmark of his efforts was the first public lecture of the HES under the title *Ευγονία. Μια Έκκλησις* (Eugenics: An Appeal) given in March of 1955 and attracted an audience of eight hundred people. Since then, the HES gained more public acceptance and respect. Moreover, they cooperated with similar associations, such as the National Union of Sanitary Education, which was the Greek department of the *Union*

International d' Education Sanitaire, an international non-governmental organisation. They organised a series of public lectures on subjects of eugenics and heredity during 1955–1956. The PIKPA, an institution for the protection of mothers and children, also collaborated with the HES. Its president, Lina Tsaldaris, and its medical director, Konstantinos Saroglou, were members both of the HES and the National Union of Sanitary Education. Furthermore, in this chapter, the mutual relationship between eugenics and politics is also discussed by putting emphasis on the participation of eugenicists in governmental positions.

In the fourth chapter, the development of an international network among the Greek, British and American eugenicists is discussed. The HES's international recognition is examined through the correspondence between the HES and its foreign contacts, mostly with the regional department dealing with Europe, the Near East and Africa of the IPPF. Correspondence and visits abroad led to the development of a profound relationship among the HES, the IPPF and other foreign associations. Maro Kanavarioti, the Secretary of the HES, was the protagonist in the developed network. Her visits to Stockholm, London and Rome were decisive to the international recognition of the HES and Greek eugenics on many levels, including the foreign press. In this context, the relationship between Clarence Gamble and Greek eugenicists is also revealed. In this respect, the fourth chapter shows Gamble's leading role in the popularisation of birth control by his collaboration with gynaecologists, either members or non-members of the HES. Furthermore, Gamble's delegates visited the country and assessed the level of family planning awareness and usage of contraceptives. Their interest in exploring the socio-medical perspective for family planning led them to reach women's associations, the PIKPA and clinics for prenatal care. In this chapter, the divergence in Louros' viewpoint for contraceptives is illustrated as well. Although he was initially a keen supporter of contraception and accepted with gratitude Gamble's offer for supplying him with contraceptives to distribute to his patients, he ended up questioning their practicality and gradually became disassociated from Gamble.

Chapter 5 is mainly devoted to the public conferences organised by the HES which, in a great extent, represent the dominant attitude of the Greeks. Taking into consideration that Greece is a small country, in terms of population, the academic and scientific elite was concentrated in Athens. Therefore, we encounter eminent scholars and scientists and sometimes politicians to be either presenters or guests at these conferences. As such, the minutes of the conferences provide a window on prevailing views. The most important issues were population problems and demography, such as overpopulation, ecology, population ageing and reproduction problems. Moreover, the fifth chapter includes the rise of feminism and its role in the change of the family model, sex and health education. As a common eugenics subject, many aspects of hereditary diseases are also included in the chapter. The aforementioned topics of discussion were drawn upon a timeline beginning from the 1950s when the problem of overpopulation, associated with ecology and population ageing, attracted much attention globally; followed by the rise of the feminist movement in Greece in the 1960s; the preoccupation with prenatal diagnosis of hereditary diseases resulting from the advances in medicine and genetics in the 1970s; and the passing of relevant legislation in the 1980s.

NOTE

1. For a detailed analysis of the history of race and racism see: Turda, Marius and Maria Sofia Quine. 2018. *Historicizing Race*. London: Bloomsbury Academic.

REFERENCES

Adams, Mark B. 1990. *The Wellborn Science: Eugenics in Germany, France, Brazil and Russia*. Oxford: Oxford University Press.

Allen, Garland E. 2001. Is a New Eugenics Afoot? *Science* 294 (5540): 59–61.

Bampiniotis, Georgios. 1998. *Dictionary of the Greek Language*. Athens: Kentro Lexicologias.

Barmpouti, Alexandra. 2015a. Population, Urbanization and Eugenics in Athens, 1950s–1970s. *Revista de Anthropologie Urbana* 5 (1): 73–81.

Barmpouti, Alexandra. 2015b. Eugenics and Induced Abortions in Post-War Greece. *Acta Historiae Medicinae, Stomatologiae, Pharmacie, Veterinae* 34 (1): 38–50.

Bashford, Alison. 2014. *Global Population. History, Geopolitics and Life on Earth*. New York: Columbia University Press.

Bashford, Alison, and Philippa Levine. 2010. *The Oxford Handbook of the History of Eugenics*. Oxford: Oxford University Press.

Burnova, Evgenia. 2016. *The Inhabitants of Athens, 1900–1960. Demography*. Athens: National and Kapodistrian University of Athens.

Cogdell, Christina. 2004. *Eugenics Design: Streamlining America in the 1930s.* Philadelphia: University of Pennsylvania Press.

Connelly, Matthew. 2008. *Fatal Misconception: The Struggle to Control World Population.* Cambridge, MA: Harvard University Press.

David, Henry P. 1992. Abortion in Europe, 1920–91: From a Public Health Perspective. *Studies in Family Planning* 23 (1): 1–22.

Descochers, Piere, and Christine Hoffbauer. 2009. The Post War Intellectual Roots of the Population Bomb. Fairfield Osborn's "Our Plundered Planet" and William Vogt's "Road to Survival" in Retrospect. *The Electronic Journal of Sustainable Development* 1 (3): 73–97.

Dowbiggin, Ian Robert. 2002. 'A Rational Coalition': Euthanasia, Eugenics and Birth Control in America, 1940–1970. *Journal of Policy History* 14 (3): 223–260.

Duster, Troy. 2003. *Backdoor to Eugenics.* New York: Routledge.

Ehrlich, Paul. 1968. *The Population Bomb.* Cutchogue: Buccaneer Books.

Ekberg, Merryn. 2007. The Old Eugenics and the New Genetics Compared. *Social History of Medicine* 20 (3): 581–593.

Ekberg, Merryn. 2013. Eugenics: Past, Present, and Future. In *Crafting Humans: From Genesis to Eugenics and Beyond*, ed. Marius Turda, 89–108. Goettinge: V&R Unipress.

Epstein, Charles J. 2003. Is Modern Genetics the New Eugenics? *Genetics in Medicine* 5: 469–475.

Galton, Francis. 1869. *Hereditary Genius: An Inquiry into Its Laws and Consequences.* London: Macmillan.

Gillette, Aaron. 2007. *Eugenics and the Nature-Nurture Debate in the Twentieth Century.* London: Palgrave Macmillan.

Glass, David. 1966. Family Planning Programs and Action in Western Europe. *Population Studies* 19 (3): 221–238.

Halkias, Alexandra. 2004. *The Empty Cradle of Democracy: Sex, Abortion and Nationalism in Greece.* Durham and London: Duke University Press.

Ioannidi-Kapolou, Elisabeth. 2004. Use of Contraception and Abortion in Greece: A Review. *Reproductive Health Matters* 12 (24): 174–183.

Kallis, Aristotelis. 2007. Racial Politics and Biomedical Totalitarianism in Interwar Europe. In *Blood and Homeland: Eugenics and Racial Nationalism in Central and Southeast Europe, 1900–1944*, ed. Marius Turda and Paul Weindling, 389–415. Budapest: Central European University Press.

Karakatsani, Despina, and Vasiliki Theodorou. 2010. *'Hygiene Imperatives': Medical Observation and Social Care of the Child During the First Decades of 20th Century.* Athens: Dionikos.

Karakatsani, Despina, and Vasiliki Theodorou. 2012. Eugenics, Childcare and Hygienic Concerns in Interwar Greece. In *Anthropological and Sociological Approaches of Health*, ed. Manos Spiridakis and Charalambos Economou, 483–510. Athens: I. Sideris.

Kline, Wendy. 2001. *Building a Better Race: Gender, Sexuality and Eugenics from the Turn of the Century to the Baby Boom*. Berkeley: University of California Press.

Kühl, Stefan. 1994. *The Nazi Connection: Eugenics, American Racism and German National Socialism*. New York: Oxford University Press.

Liddell, Henry George, and Robert Scott. 1940. *A Greek-English Lexicon*. Oxford: Clarendon Press.

Official Government Gazette. 1952. Law 2032: For the Establishment of the Hygiene Education Service Under the General Office for Hygiene at the Ministry of Social Care, 77.

Osborn, Fairfield. 1948. *Our Plundered Planet*. New York: Pyramid Publications.

Osborn, Fairfield. 1962. *Our Crowded Planet: Essays on the Pressures of Population*. Garden City, NY: Doubleday.

Paxson, Heather. 2002. Rationalizing Sex: Family Planning and the Making of Modern Lovers in Urban Greece. *American Ethnologist* 29 (2): 307–334.

Paxson, Heather. 2004. *Making Modern Mothers: Ethics and Family Planning in Urban Greece*. California: University of California Press.

Pichot, André. 2000. *La société pure. De Darwin à Hitler*. Paris: Flammarion.

Pine, Lisa. 1997. *Nazi Family Policy, 1933–1945*. London: Bloomsbury.

Plato. 1948. *The Republic*, trans. L.A. Dunlop. London: J.M. Dent & Sons Ltd.

Proctor, Robert N. 1998. *Racial Hygiene: Medicine Under the Nazis*. Cambridge, MA: Harvard University Press.

Ramsden, Edmund. 2002. Carving Up Population Science: Eugenics, Demography and the Controversy Over the "Biological Law" of Population Growth. *Social Studies of Science* 32 (5–6): 857–899.

Ramsden, Edmund. 2009. Confronting the Stigma of Eugenics: Genetics, Demography and the Problems of Population. *Social Studies of Science* 39 (6): 853–884.

Roukas, Constantinos. 1979. *Sexual Intercourse and Induced Abortion Rates of Students in Athens*. Athens: Laboratory of Hygiene and Epidemiology, University of Athens.

Soloway, Richard A. 1995. *Demography and Degeneration: Eugenics and the Declining Birthrate in Twentieth-Century Britain*. Chapel Hill: University of North Carolina Press.

Stepan, Nanvy Lays. 1991. *'The Hour of Eugenics': Race, Gender and Nation in Latin America*. Ithaca, NY: Cornell University Press.

Stern, Alexandra Minna. 2005. *Eugenic Nation: Faults and Frontiers of Better Breeding in Modern America*. Berkeley: University of California Press.

Trubeta, Sevasti. 2007. Anthropological Discourse and Eugenics in Interwar Greece. In *Blood and Homeland: Eugenics and Racial Nationalism in Central and Southeast Europe 1900–1940*, ed. Marius Turda and Paul Weindling, 123–144. Budapest: CEU Press.

Trubeta, Sevasti. 2010. The Strong Nucleus of the Greek Race: Racial Nationalism and Anthropological Science. *Focaal-Journal of Global and Historical Anthropology* 58: 63–78.

Trubeta, Sevasti. 2013. *Physical Anthropology Race and Eugenics in Greece (1880–1970s)*. Leiden: Brill.

Tseperi, Popi, and Elisabeth Mestheneos. 1994. Paradoxes in the Cost of Family Planning in Greece. *Planned Parenthood in Europe* 23 (1): 14.

Turda, Marius. 2010a. Whither Race? Physical Anthropology in Post-1945 Central and Southeastern Europe. *Focaal-Journal of Global and Historical Anthropology* 58: 3–15.

Turda, Marius. 2010b. *Modernism and Eugenics*. London: Palgrave Macmillan.

Turda, Marius. 2010c. Race, Science and Eugenics in the Twentieth Century. In *The Oxford Handbook of the History of Eugenics*, ed. Alison Bashford and Philippa Levine, 62–79. Oxford: Oxford University Press.

Turda, Marius, and Aaron Gillette. 2014. *Latin Eugenics in Comparative Perspective*. London: Bloomsbury.

Turda, Marius, and Maria Sofia Quine. 2018. *Historicizing Race*. London: Bloomsbury Academic.

Turda, Marius, and Paul Weindling. 2007. *Blood and Homeland: Eugenics and Racial Nationalism in Central and Southeast Europe, 1900–1944*. Budapest: Central European University Press.

United Nations. 1948a. *Convention on the Prevention and Punishment of the Crime of Genocide*. Resolution 260A.

United Nations. 1948b. *Universal Declaration of Human Rights*. Resolution 217A.

Valaoras, Vasilios. 1969. *The Sub-Fertility of the Greeks and Induced Abortions*. Athens: n.p.

Valaoras, Vasilios, Antonia Polychronopoulou, and Dimitrios Trichopoulos. 1969. Greece: Postwar Abortion Experience. *Studies in Family Planning* 46 (1): 10–16.

Vogt, William. 1948. *Road to Survival*. New York: William Sloane Associates Inc.

Weindling, Paul. 1989. *Health, Race and German Politics Between National Unification and Nazism, 1870–1945*. Cambridge: Cambridge University Press.

Weindling, Paul. 2004. *Nazi Medicine and the Nuremberg Trials: From Medical War Crimes to Informed Consent*. London: Palgrave.

Zero Population Growth. 1972. Dir. Michael Campus. USA: Sagittarius Productions Inc.

Ziegler, Mary. 2008. Reinventing Eugenics: Reproductive Choice and Law Reform After the World War II. *Cardozo Journal of Law and Gender* 14: 319–347.

CHAPTER 2

Health and Hygiene, 1900–1950

1 Historical Background and Demography

During the first half of the twentieth century, war and disease dominated in Europe, including Greece. The country experienced many years of consecutive wars which impacted to the health of the population. Unfortunately, the Greek state often adopted unsuccessful and inadequate hygienic measures, because the conditions were rarely favourable, alongside financial limitations.

The Greek Revolution against the Ottoman rule reached its peak in 1821 when about one-third of the country became independent. Afterwards, more revolutions followed albeit unsuccessful. The beginning of the century was marked by a historical defeat by the Ottoman Empire in 1897. The First Balkan War took place few years later in 1912–1913. Luckily for Greece, by the end of the First Balkan War, the country managed to annex another large part of its territory. A second annexation of the 'old land' occurred by the end of the Second Balkan War and the Treaty of Bucharest (1913). Then, Greece acquired the prefectures of Macedonia, southern Epirus and the island of Crete but territories, such as Thrace and the Dodecanese islands were still under foreign rule. Although the Balkan Wars resulted in the addition of a large part of territory and population to the country, the First World War followed few months later bringing serious epidemics and high mortality to the local population. War and disease went hand-in-hand during the

© The Author(s) 2019
A. Barmpouti, *Post-War Eugenics, Reproductive Choices and Population Policies in Greece, 1950s–1980s*,
https://doi.org/10.1007/978-3-030-03568-6_2

first half of the twentieth century. Continuous war conflicts, poverty and unhealthy living conditions facilitated the spread of infectious diseases, such as cholera, malaria and typhus fever. In particular, in the region of Eastern Macedonia, where Greek, British, French and Serbian soldiers coexisted, the diseases eradicated hundreds of soldiers and local civilians.

At the same time, similar to the rest of the Southeast European countries, such as Bulgaria, Yugoslavia and Romania, the Greek public health system was practically non-existent before the 1920s, although the relevant legislation did exist. The Balkan Wars and the First World War, as well as the disastrous warfare in Asia Minor in 1922, had negative consequences for the general health of the population (Hirschon 2003). Nevertheless, private initiatives by individuals like Konstantinos Savvas and Emmanuel Lambadarios were decisive for the reform of public health in Greece during the 1920s. There was an obvious duality in the role of the physician, who was concerned more with society in general than the individual solely, trying to connect individual physical health with morality, and public health with social norms. A leading figure of the hygienic movement was Konstantinos Savvas. Long before, his classic handbook of hygiene was published (Savvas 1928); he took many initiatives for the protection of the population. Indicatively, as early as 1905, he organised a hygienic movement, the 'Anti-malaria League' (Σύλλογος προς περιστολήν των ελειογενών νόσων). The first attempt to eliminate the disease occurred in 1908, when a law, providing for the free distribution of quinine, the medication for tuberculosis and malaria, was passed (Official Government Gazette 1908a). A year later, in 1909, the First National Conference on Tuberculosis took place in Athens (Gardikas 2018). Moreover, during the Balkan Wars there was a coordinated effort, under the direction of Savvas, to eradicate disease among Greek soldiers. Savvas made great efforts to help the Greek soldiers on the battlefields, while also protecting the population of Northern Greece, where the fighting took place and many diseases were endemic.

In 1910, Savvas and Lambadarios proposed the first complete plan for the reform of sanitary services and the supervision of public health, which gained parliamentary approval. Primarily, it concerned the protection from contagious diseases and the regeneration and healthy reproduction of the race. The plan was aimed at the reconstruction of health institutions and policies in order to protect mothers and children from conception until school age. They claimed that the quality of children's health was crucial for the biological quality of the race; the state was thus obliged to provide the best conditions for mothers.[1] For instance, the existing Law 4029 for the labour of women and minors, introduced in 1912, already stipulated that placing heavy labour on children led to the feebleness of their body and mind, unavoidably leading to an unhealthy population (Official Government Gazette 1912). As Karakatsani and Theodorou argued, this view echoed wider European developments towards the protection of children from hard labour in factories, in addition to establishing obligatory education and passing laws for the protection of children (Karakatsani and Theodorou 2010). The rationale was that children who were born and raised under optimal conditions would renew and strengthen the nation's human capital. For a long period, the principal concern of the Greek state was the creation of a durable army. Although the health of the population was in decline, there were no important initiatives towards its protection. During the period from the late nineteenth century until 1914, the public health sector was very poor and lacking proper infrastructure. In this context, many health institutions and hospitals were obliged to cease operation (Dardavesis 2008). The same view prevailed during the post-war period when memories of the Second World War were fresh.

The continuous warfare resulted in the addition of territory and a large number of immigrants but at the same time the country suffered from loss of predominantly male population in the battlefields, the uncontrollable spread of epidemics, famine and poverty. In addition, the majority of population was gathered in urban centres causing health and hygiene side effects. The density of inhabitants in the wider region of Attica, including both Athens and Piraeus, facilitated the spread of diseases because it was practically impossible for the state to protect its citizens under these circumstances.

2   The Public Health System

Political polarity and instability throughout the first half of the century impacted negatively on the development of the public healthcare system and exacerbated the already poor sanitary living conditions. Only a radical reform of the public health system could offer viable solutions, which meant the adoption of new regulations. However, during the short peaceful periods, the collaboration of the Greek government with international organisations produced remarkable results in terms of health and hygiene policies and medical training. Tellingly, during the inter-war years, the liberal government of Eleftherios Venizelos improved the health sector as the Prime Minister was also the Minister of Health. The government put emphasis on the improvement of public health services.

In this context, the most significant action of the Greek state was the passing of Law 346 for the unification of public health services into one central hygiene service introduced in 1915 (Official Government Gazette 1915). The law included detailed description of the new organisation of public hygiene, including required qualifications of the personnel at hygiene services, the duties of the Medical Council (*Ιατροσυνέδριον*) and the Inspector of Hygiene; the role of the prefectural medical officer (*Νομίατρος*), the duties of the personnel at quarantine hospitals (*λοιμοκαθαρτήρια*), the duties of the representatives of public hygiene abroad, the regulation of vaccinations, the duties of those who collected dead bodies, and finally, the amount of taxes for quarantine hospitals. This initiative became a significant factor of the reformation of the public health sector.

Until then, the most important hygiene service was the Medical Council, which was founded in 1834 (Official Government Gazette 1834), as a part of the Secretariat of the Ministry of the Interior. At the time all public health services resided at the Ministry of the Interior. The Medical Council was composed of a president and six members, four physicians and two pharmacologists. Later, one or two veterinari-ans were added. The main duty of the Medical Council was to inspect the work of physicians, surgeons, dentists, veterinarians, pharmacologists and midwives. Secondly, the council was responsible for undertaking issues of medical jurisprudence. Thirdly, it was the official consultant of the Secretariat of Interiors for any medical matter. As it was the advisory board on every matter of health and hygiene, its members and its work were highly respected. Savvas, for instance, was President of the Medical Council from 1897 to 1908.

In 1917, the health sector was separated from the Ministry of the Interior in order to form a separate Ministry of Social Care (Official Government Gazette 1917). In 1920, Savvas made continuous efforts for its reform and its change into the Ministry of Hygiene and Social Care. Law 2882, which included Savvas' proposals for the improve-ment of public health and hygiene, was indeed passed by the Third Constitutional Assembly, but never implemented, because of the military catastrophe in Asia Minor and its tragic consequences two years later.

The wave of refugees from Asia Minor to mainland Greece was the landmark for the reform of the public health system, mainly because of overpopulation and the uncontrolled transmission of diseases. Due to this unexpected growth in the population, there was an immediate need for new health policies. It was then that the inadequacies of the public health system came to light because it was unable to respond effectively to such urgency. Thus, the need to protect and help the citizens became the first priority. Hence, in 1922 the Ministry of Social Care was incorporated into the new Ministry of Hygiene, Care and Perception (Official Government Gazette 1922). For the next four years, the ministry was organised and developed according to novel legal enactments, which defined its services.

Support received from international organisations was also essential. In 1923, the Epidemic Commission of the League of Nations Health Organisation (LNHO) visited Greece, helped the sanitary organisation of refugee camps and undertook preventive vaccinations. Institutions like LNHO were manned by health experts who worked mainly on research and eradication of epidemics, like malaria, tuberculosis and leprosy. The LNHO collaborated with the Rockefeller Foundation (Weindling 1997), which was also very active in Greece during the 1930s and 1940s, especially with the anti-malaria campaign. In 1928, health experts from LNHO conducted research on malaria and tuberculosis and offered their findings and advice to the Greek government. Their contribution helped the reorganisation of public health policies, by improving the organisation of the public health system and introducing local physicians to international standards of hygiene, sanitary housing and nutrition (Karakatsani 2011).

Yet, for a short period, during the dictatorship of General George Pangalos (1925–1926), the Ministry of Hygiene was abolished and its services were allocated to the Ministries of the Interior, Education and Military (Official Government Gazette 1926a). Once more, political instability disrupted the organisation of the public health sector. However, it was during Pangalos' government when the law for the protection of the children until the age of two as well as their mother's protection by the state was implemented. The objective of this action was to tackle the growing problems of infant mortality, abortion and abandonment. A few months later, the government of Georgios Kondylis re-established the Ministry of Hygiene, Care and Perception and added the Secretariat of Hygiene (Official Government Gazette 1926b).

Once more, during the 1930s the presence of infectious diseases motivated the various governments to pay close attention to the level of hygiene among the population and adopt sanitary measures, particularly with respect to the prevention of tuberculosis and malaria, which were endemic in Greece (Theodorou 2002). In 1928, a dengue fever epidemic ravaged a large part of the population too (Kiriopoulos 2008). During the period under examination, the battle against tuberculosis and other diseases became more active and effective. Activities and initiatives like the organisation of open-air camps and schools; the reorganisation of the Sotiria Sanatorium in Athens, the biggest in Greece; the founding of more sanatoria; as well as many preventive medical examinations and vaccinations in schools were some examples of the methods to eradicate infectious diseases (Stoyiannidis 2016). During this period, it was due to private initiatives that hospitals and sanatoria were established in Greece. The health of the Greeks rested exclusively in the hands of wealthy individuals. Many groups—mostly of wealthy women—run errands in order to help the poor and the needy. The spirit of solidarity and charity was truly remarkable but eventually inadequate. The need to build a public health system was beyond dispute.

In 1929, during the last period of the Liberal government under the Prime Minister Venizelos, the Ministry of Hygiene, Care and Perception was renamed again as the Ministry of Hygiene (Official Government Gazette 1929a). Moreover, an important step taken by the Liberal government towards the protection of mothers and children was the creation of a scientific committee in the Ministry of Hygiene to supervise all institutions associated with the protection of mothers and children. In addition, the government contributed financially towards the activities organised by the Patriotic Institution of Healthcare (Πατριωτικό Ίδρυμα Περιθάλψεως), later the PIKPA. Generally, the government's objective was to gradually replace the private charity funds that had undertaken the health care of the nation with state funding. During the same period, they created a special school for children with tuberculosis. These children were categorised in five groups, according to their mental and physical state. They were: very thin; mentally distorted; foreign language speakers; illiterate and working.

Next to the political stability of the period 1928–1932, the fact that Venizelos undertook the responsibilities of the Ministry of Health contributed to the high level of effectiveness of hygienic measures (Karakatsani and Theodorou 2010). In 1932, the Departments of Hygiene and

Health care were unified and formed the Ministry of State Hygiene and Perception (Υπουργείο Κρατικής Υγιεινής και Αντιλήψεως).

A year earlier, the School of Hygiene was established in Athens (Official Government Gazette 1929b). Similar schools were established in other countries at the same period, in London (1924), Zagreb (1928) and Ankara (1936), all with the support of the Rockefeller Foundation (Dardavesis 2005). The school's purpose was epidemiological research and education, research on the impacts and effectiveness of medication against diseases and theoretical and practical teaching of malarial diseases. It was an institution for higher education and the first to offer special-isation in hygiene. Norman White, the representative of the League of Nations in Greece, was the first Director of the School. From the begin-ning of its activity, a group of experts belonging to the Rockefeller Foundation was established in the school, contributing to both the edu-cational work and the anti-malaria campaign (Gardikas 2008). Among them were M. Balfour, M. Barber, J. B. Rice, R. C. Shannon and D. E. Wright. The Rockefeller Foundation also offered scholarships for overseas training. The contribution of the Rockefeller experts to the anti-malarial campaign was significant; and despite the fact that they left Greece in 1938, the campaign was continued by Greek experts who they had trained.

Alexander Koryzis, Minister of Health during the government of Ioannis Metaxas (1936–1941), appointed a committee under the direc-tion of Fokion Kopanaris, for the comprehensive study of malaria with the purpose of finding effective ways for its eradication. A combination of specialised personnel, sufficient funding and major drainage works in Northern Greece, Thessaly and Epirus, made the work of this com-mittee very successful. However, a second wave of malaria incidents fol-lowed the famine outbreak in 1942 (Hionidou 2006). Therefore, the eradication of malaria was claimed later than the aforementioned effort, during the 1950s and 1960s. Daniel Wright, for instance, returned to Greece as a Director of the United Nations Relief and Rehabilitation Administration Medical Division Mission to supervise the country's anti-malaria program. After the Second World War, he supervised the DDT spraying in the country (1946), a radical method of eradicating mosquitoes, which transmit the disease between people (Gardikas 2018).

Laws 5733/1932 introduced by Venizelos' government and 6298/1934 introduced by Tsaldaris' government regarding the Institution of Social Insurance (IKA) contributed to the strengthening of social welfare and security in Greece. Funded partly by employers and partly by workers, the IKA would, respectively, offer pension for the aged and insurance in sickness. Indeed, the state's contribution to the IKA was just the management of its budget, since it could not offer financial aid (Nikolaidis and Sakellaropoulos 2002). Nonetheless, these laws were implemented only during the dictatorship of General Ioannis Metaxas when he tried to establish a programme for social care in favour of the lower classes. In this context, the government passed a series of laws to accomplish this. Among them, the Law 965 regarding the organisation of public health institutions and hospitals (Official Government Gazette 1937a) and the Law 547 for the eight-hour workday (Official Government Gazette 1937b) were the most significant. As in the last period of the Liberal government, the political stability of the period 1937–1940 benefited the implementation of new legislation for public health care. Notwithstanding, its application was often restricted due to limited public funds (Sarantis 2009).

3   School Hygiene

In general, during the pre-war period public health policies in Greece focused on children because this target group was regarded as an investment in future citizens and soldiers. In turn, the state focused not only on their protection, but also on their health improvement. Implementation of preventive medicine had been the state's priority, due to its effectiveness, in terms of both health improvement and cost.

Responding to the demand for children protection, Emmanuel Lambadarios was also the leading person who popularised the pedological movement in Greece. Pedology offered the theoretical concept of building a new health system directed towards the protection of children. The science of pedology was introduced in Greece by Lambadarios in the beginning of the twentieth century and by 1936 it became a university course at the University of Athens. Lambadarios also founded the Pedological Institute and during the 1910s founded children's camps, student polyclinics, with the first student polyclinic to be founded in Athens in 1915, and open-air schools for pupils who were susceptible to tuberculosis. Such works of social perception were usually funded by private organisations, whose contribution was vital for the development of the public healthcare sector (Stoyiannidis 2016). In 1920, the journal *Pedology* (*Παιδολογία*) and in 1936 the journal *School Hygiene* (*Σχολική Υγιεινή*) were published as a means of disseminating pedology.

Apart from physicians, scholars from other scientific branches were interested in the protection of children's health. One of them, for instance, was Nikolaos Exarchopoulos, a pedagogue and supporter of experimental pedagogy who became President of the Academy of Athens in 1942. He argued that it was almost impossible to separate the scientific fields of pedology and pedagogy. Pedology approached childhood theoretically, whereas pedagogy was more practical (Exarchopoulos 1950). In order to justify the association of pedagogy with practical sciences, Exarchopoulos also described its connection with biology. He argued that although the contribution of the advances and discoveries of biology was important, at the same time they were limited to the biological side of the individual. Instead, pedagogy regarded and researched the child holistically. Notwithstanding, Exarchopoulos admitted that evolutionary biology directed all pedagogical research, because it permitted the proper education of children according to heredity, fitness and physical development. Indeed, his studies of children were inextricably linked to certain eugenic practices, such as adhesion to the mathematical interpretation of the individual and its classification according to its proximity to 'normality'.

Equally important, Exarchopoulos founded the Experimental Laboratory (*Πειραματικό Εργαστήριο του Πανεπιστημίου Αθηνών*) in 1923 and the Experimental School (*Πειραματικό Σχολείο του Πανεπιστημίου Αθηνών*) in 1929, both belonging to the University of Athens. Exarchopoulos also founded the Psychological Laboratory of Athens (*Ψυχολογικό Εργαστήριο Αθηνών*) and the Psychological Laboratory of Thessaloniki (*Ψυχολογικό Εργαστήριο Θεσσαλονίκης*). The Experimental Laboratory aimed at introducing the practical pedagogical methods to students of pedagogy. Its target was the holistic research of Greek pupils from the physical, psychological and moral viewpoints. As far as physical research was concerned, the projects focused on the biological development of Greek pupils. To this end, they used a variety of special tools to define accurate anthropometric dimensions of pupils. They investigated and registered height, weight, thorax perimeter, head diameter and muscle strength, in order to specify, the level of normality at each age and underline the differences between the sexes and social classes. Another of Exarchopoulos' important studies was to compare Greek children to those of other nations (Exarchopoulos 1950). Biometry was one of eugenics' methodologies, widely practised at the beginning of the twentieth century, in combination with Mendelism and pedigree studies. Biometry was used to prove the hereditarian nature of a trait or behaviour. One of the most renowned examples of institutionalised research on biometrics was the Galton Laboratory at the University College in London (Mazumdar 1992; Jensen 2002). Undoubtedly, Exarchopoulos was inspired by its research.

As far as the psychological research was concerned, the students of the laboratory researched the intelligence level, the differences among social classes, between sexes, and drew comparisons with pupils of other countries. Moreover, they registered the consistency of teachers' work and pupils' perceptiveness, acuity and critical ability. Furthermore, they investigated the level of attention, concentration, fitness and the familial influence on a pupil's intelligence. In addition, it is important to mention that Exarchopoulos created the Greek version of the Binet-Simon I.Q. test (Exarchopoulos 1931).

The Experimental School was the place where the above-mentioned studies took place. They used psychographs and medical records of the pupils in order to create indexes of the children's performance. Based on these studies, they characterised them as uppermost, inferior or mediocre. The classification of pupils was a common phenomenon, made either by pedagogues and teachers or by school doctors, because pedology and pedagogy were also linked with school hygiene. All three scientific branches researched and endeavoured to improve children's mental and physical health and intelligence. Nonetheless, their practices often crossed the border between health improvement and positive eugenics.

School hygiene was part of the public hygiene, but it was particularly favoured by the Greek state. The teaching of hygiene practices in schools was used to implement wider ideas and practices of hygiene in every household. Children, who would acquire hygienic knowledge and attitudes at school, would then carry them home and so affect the attitude of the rest of the family. In the long run, pupils with better hygienic attitudes would become stronger workers, forceful soldiers and healthier people, who would produce future healthy families. The first Office of School Hygiene, which organised activities regarding school hygiene, was founded in 1908 and its first Director was Georgios Drosinis (Official Government Gazette 1908b).

School hygiene was concerned with two areas: on the one hand, with school buildings and on the other hand, with teachers and pupils. School premises were populated areas where infectious diseases could be easily transmitted. School buildings were to be built in accordance with the basic rules of hygiene; thus, they should be clean, airy and sunny. Furthermore, access to the school premises was prohibited to pupils or teachers who lived at the same house with someone suffering from a contagious disease or prone to such a disease, like tuberculosis. In this context, an excellent example for the prevention of diseases was the organisation of open-air camps and open-air schools, introduced by Lambadarios. Children who were prone to tuberculosis benefited from those open spaces, where they could both be educated and amused. Most of the times, monasteries offered the ideal environment to be used as open-air spaces for children.

School hygiene was based on the work of school doctors who were physicians who specialised in school hygiene and had at least three years of experience (Official Government Gazette 1914). Their duties included the medical treatment of pupils; promotion of preventive hygiene; and isolation of the sick from the healthy. They were responsible for supervising the building, checking the teaching methods, examining and vaccinating pupils. In fact, school doctors were more responsible for preventive than curative medicine. School doctors were to be attentive and vigilant with the patients and their examination results. According to Law 240, they were allowed to take research leave to travel to Western Europe to learn new methods of school hygiene. Countries like Belgium and Germany were already experienced in the activities of school doctors, as they had appointed them at their schools much earlier than Greece had.

Moreover, the work of the school doctor was aided by school nurses and assistants. The role of the school nurse was equally as essential as that of the doctor, mediating between the doctor, the pupil and the pupil's family. School nurses visited sick pupils at their own houses and built up a relationship with their family. Usually, female school nurses could approach the pupils' mothers much easier than the doctor. As a result, they could educate them about hygiene. Therefore, school nurses played a more important role outside the school than inside it. During the period when Lambadarios was Director of the Service for School Hygiene at the Ministry of Public Education, he appointed 15 inspectors of school hygiene, 70 school doctors and many more school nurses. However, during the period between 1926 and 1933, due to limited public funds, the Service for School Medicine was abolished and only 20 school doctors were working around the country (Papaioannou 1939).

Along with the examination, school doctors filled up the newly introduced personal health card of each pupil (ατομικόν δελτίον υγείας μαθητού). The physician examined the pupil both physically and mentally and registered the results on this card. The process was repeated frequently, in order to register and monitor the progress of the child. Each pupil's records were kept until the age of eighteen. Papaioannou's work, *Student's Health Card*, offered the most detailed analysis of the purpose and the use of student health cards.

Primarily, Papaioannou highlighted the dangers of childhood, such as childhood diseases, abnormal development and bad schooling conditions. School hygiene, in general, and the health card, in particular, aimed at the elimination of these degenerative factors. Furthermore, contributing factors over the course of life of every pupil which needed attention by the school doctor included: family life, school life, housing and nutrition. The harsh living conditions of the period under examination were depicted in the health of the population, most notably that of vulnerable pupils. The health card was indicative of their physical and psychological state. This record was often regarded as a means of preventive medicine. As already mentioned, school doctors examined their pupils frequently, in order to keep a record of their development. The continuity of the results offered the possibility of predicting the state of health of the examined person or to prevent the spread of a disease. In this way, school doctors were alerted to an undesirable result and sought for ways to improve the physical and mental health of the child.

As far as the actual examination was concerned, parents participated as well. They were present during the examination, and they had access to the results and the health card. Furthermore, parents were asked for the medical history of the family. Their presence was crucial, because it permitted school doctors to obtain a better image of the pupil's health. Regarding the family, the rest of the members could be protected from a latent disease or a variation from normality detected in the pupil. At that point, Papaioannou underlined the usefulness of the health card regarding protection against malaria.

By 1920, school doctors used a variety of tools to measure the physical characteristics and dimensions of the pupils. They measured the head, thorax, height, weight, etc. Based on the statistics of their findings, they could assume the 'factor of robustness' and estimate the 'average Greek pupil'. Karakatsani and Theodorou argued that these practices established which pupils were 'eugenic' and which were 'dysgenic' (Economou and Spyridakis 2012). Similar assumptions could be reasonably argued due to the mathematical nature of the examination. It was unavoidable to find the average measurement and compare it with the rest. According to Papaioannou, pupils were categorised in three categories; healthy, under surveillance and sick or under treatment. As previously shown child classification continued to be used in experimental pedagogy (1923) and in the planning of health policies, promised by the last Liberal government (1928–1932).

Apart from being a preventive measure, the student heath card served as a way to evaluate the results of theoretical and physical education. On the one hand, there was a series of measurements, records and statistics for each pupil, while on the other hand closer observation and comparison among health cards revealed the condition of each school as a whole. The role of pedagogy was to gather those statistical facts, in order to evaluate its own work based on those findings and use proper guidelines to construct a forward-thinking, more effective educational system. As a consequence, there was a mutual and significant relationship between pedagogy and school hygiene.

As already mentioned, ideas about preventive medicine often led to positive eugenic proposals. Papaioannou, for instance, was one of those who supported state intervention in public health by examining the health of Greek people at a larger scale, not only in schools. He argued that the possibility of issuing a health card in many public sectors simultaneously and continuously could solve racial and national problems. Health cards at schools, military camps and workplaces would aid the creation of family trees and the advance of racial research. He obviously admired and endorsed Galton's ideas. Although he briefly described the condition in other European countries, he particularly praised Britain. He attributed the success of school hygiene in Britain to the dissemination of eugenics and the work of Galton and his laboratory. It was obvious that he would have liked to apply the same eugenic methods in Greece to fight racial degeneration. Papaioannou was one of many Greek scholars who were familiar with eugenics ideas. The fact that they were aware of eugenics even from the beginning of the twentieth century is depicted in their publications. Among them, Moisis Moisidis was the most productive in literature in the 1920s (Moisidis 1922, 1925a, b, 1928).

4   EUGENIC LITERATURE

The limits of state intervention in the life of individuals in terms of promoting public hygiene were a matter of concern of scholars, physicians and politicians. As in many European countries, eugenics reached Greece in the beginning of the twentieth century. Greek scholars and physicians who studied or worked abroad spread the word of Sir Francis Galton in Greece and soon eugenics gained supporters. At the time, the Greeks were mostly attracted by racial protectionism. Therefore, it was not surprising that attention was paid to the mental and physical qualities of prospective parents. In this context, eugenicists suggested state intervention in married couples and families.

Apostolos Doxiadis, Minister of Healthcare during 1922–1928, elected Senator in 1932 and President of the Patriotic Institution of Social Welfare and Awareness (PIKPA) during 1924–1932. He was the father of Spyros Doxiadis, a renowned professor of paediatrics, President of the Hellenic Eugenics Society and Minister of Health. According to A. Doxiadis, it was the state's obligation to intervene in families in order to enhance the biological value of the race. Therefore, state intervention in family matters was unavoidable. He argued that every family should have on average four children, provided that it had the ability to raise them properly in a hygienic environment, at least until the age of five. Similarly to Papaioannou, A. Doxiadis suggested that every family retain a record of births, congenital diseases, bad habits, such as alcoholism and drug addiction, profession and education of each member. These details would allow a biological evaluation of the family by the state. At this point, the state would decide whether to encourage or discourage this family from reproduction. One of his significant suggestions was that the state should financially aid those poor families which had high biological value. A. Doxiadis did not associate biological quality with social class. On the contrary, he acknowledged the possible biological value in every person or family regardless of their financial state. In addition, he claimed that the reconstruction of society should be done on the basis of race, not of social class. A. Doxiadis did not agree with the widely supported viewpoint that low social class was associated with low biological value or poor mental capacity. At this time, it was unlikely for a eugenicist to have a broader conception of the origin of intelligence apart from a combination of heredity, high social class, proper education and good nutrition. However, A. Doxiadis believed that poor and uneducated families were valuable; therefore, he did not hesitate to propose additional taxes on unmarried individuals. In accordance with A. Doxiadis, Makridis another eugenicist also proposed to legalise a tax on the unmarried.

Among the suggestions for state intervention in family matters, Greek eugenicists expressed their disapproval of mixed marriages. For example,

Makridis argued that there should be a strict prohibition of marriage between Greeks and foreigners, which was up to that point valid only for soldiers and officers. Given that the Greek race was superior to other races, Makridis argued that there was no gain from racial mixes. According to Makridis, only inferior races benefit from mixes because mixing with a superior race would elevate the inferior's value. Added to this assumption, Makridis claimed that the Greek race has already been mixed with inferior races which were profited by the Greek race (Makridis 1940). It was obvious that he shared Ioannis Koumaris' ideology about the superiority of Greek race (Trubeta 2013). Above all, it was believed that miscegenation would undermine the quality of the Greek race. There were more cases which enforced marriage prohibition, such as marriage among relatives (incest marriages) up to the fifth grade or marriage between spouses who had an age difference of more than ten years. Makridis' advice to young people to prefer brown-haired, because fair-haired were, allegedly, more prone to tuberculosis, was also provocative. In this context, Savvas, for example, proposed the prohibition of marriage of women having deformed pelvises, because delivery would be very hard or impossible. Marriage prohibition has always been a strict eugenic measure because underminded personal freedom and human rights. However, for some eugenicists it was perfectly accepted because, in the long run, avoiding unhealthy marriages would improve society and in the case of mixed marriages, their prohibition would safeguard racial purity. To some considerable extent, the premarital health certificate was a means of managing marriages. However, it did not exert any influence because on the one hand it did not eventually prohibit the marriage or reproduction of the examined individuals and on the other hand it was not compulsory.[2]

Additionally, during the first half of the twentieth century, there was lack of information about the method of transmission of infectious and venereal diseases. Therefore, people who suffered from diseases, which were not transmissible by sexual intercourse, they might be excluded from marriage because they could allegedly transmit the disease to their spouse. In this context, Economopoulos suggested the compulsory teaching of the medical details of tuberculosis in schools, professional schools and the army. Moreover, he underlined the necessity of the declaration of tuberculosis incidents and the compulsory hospitalisation of dangerous cases. Venereal diseases, he claimed, were a danger to society and race because they caused population decline and birth defects. He proposed founding special health centres for free preventive examination as well as for the compulsory reporting of incidents by physicians (Economopoulos 1922). Furthermore, Savvas shared his view for the founding of these special centres at each hospital, where examination and medication would be free of charge. He agreed with the compulsory declaration of an infectious disease and the legal punishment of spouses who hid it. At the same time, doctors' confidentiality was also compulsory. Physicians were legally obliged to protect the anonymity of the patient, but also to declare any incidence of an infectious disease to the appropriate state authority, usually the most proximate hygienic centre. Physicians, who did not act thus, were to be punished. The archives of the Athens Medical Association record that physicians were punished for similar cases during the post-war period too (Athens Medical Association Archive 1952).

Ideas, such as Makridis', were representative of physicians who overemphasised the protection of family and procreation due to racial protectionism and eugenics ideology. He thus constructed a plan of action, including specific interventions by the state, in order to facilitate the creation of robust Greek families. First of all, he regarded the reinforcement of the institution of marriage as a priority. In addition, the state should implement policies for the protection of pregnant women and the facilitation of workplaces. Facing a rise in the number of induced abortions, Makridis suggested the need to organise the fight against abortions and abandonment of newborns. The great number of abortions became a matter of demographic concern. At this point, Makridis criticised Malthus' theory of population, because it undermined population growth and, therefore, favoured the practice of abortions.

Regarding children's health, Makridis supported the close observation of the development of children from their conception until the eighteenth year. According to him, during this period children should be educated and examined by school doctors, as previously described. Furthermore, he proposed a plan for the protection of every Greek woman who faced problems with procreation and sterility. The state should also care for and help women who needed an operation or treatment to deal with sterility. Moreover, in cases of women who did not wish to have children due to poverty, state authorities would offer financial aid, because it should not be overlooked that those women could give birth to future soldiers, workmen and citizens. At that point, Makridis praised the Finnish mother, wife and housewife for her strength and courage to keep her house in good condition, to care about her husband and raise her children with the traditions of her race. Makridis evaluated the priorities of the state according to their importance towards race regeneration; firstly, the protection of mothers, secondly of children and thirdly of families with many members.

During the first half of the twentieth century, women maintained the traditional model of mother and housewife. The role of women as individuals having free will and action was far from reality. Considering this situation, it was obvious that the above-mentioned proposals for racial improvement regarded women as a necessary component of procreation. Despite the fact that some women worked outside the house, the role of mother was always foremost. Therefore, Makridis, Economopoulos and others argued that it was very important for the mother to stay at home at least during her pregnancy and until the newborn became six months old, in order to breast-feed it. The need for absence from work for a period before and after labour was also emphasised. Economopoulos stressed the need for a public service solely dealing with maternal, newborn and infant care.

Apart from purely practical solutions to population problems, such as the health card, there was great concern about informing the public about a hygienic and healthy lifestyle. Target groups were mostly couples about to get married and pregnant women. Briefly, the state aimed at altering the lifestyle of people before marriage, during pregnancy and after birth. Sanitary conditions were so harsh that they prohibited population increase both in quantity and quality.

In this context, scholars and politicians argued that eugenics could be a means to cultivate the 'procreation instinct', which would be more effective than any other eugenic policy. Future parents should be aware of the consequences of their decisions regarding procreation. If everyone thought responsibly about future generations, they would have chosen their partner according to his/her health condition and biological value. Furthermore, A. Doxiadis had a similar view on the subject and for the first time mentioned the need to cultivate 'biological consciousness', the feeling of biological obligation of the individual to the community (Doxiadis and Fragkou, n.d.). As already mentioned, he proposed the use of a booklet, which would contain the medical history of every citizen, not only of pupils. More specifically, A. Doxiadis argued that it was very difficult to know the medical history of a family because people would hide information for the sake of marriage. The only solution that would protect the future generations was to instil in the mind of young people the obligation to care about their children. The best way to achieve such a goal would be to modify their mentality. In particular, he pointed out that eugenics should become like a new religion and the efforts of the state, the society and the family should have one goal, namely eugenics, the improvement or at least not worsening of the human race (Doxiadis and Fragkou, n.d.). It was essential to inform lay people about the potential dangers or benefits from their choice of spouses.

In accordance with A. Doxiadis, Makridis used the theory developed by Karl Pearson to argue about the possibility of achieving good quality of births by proper choice of spouse to secure proper genetic predisposition (Pearson 1909, 1912). Human enhancement in two or three generations could be achieved by the combination of proper choice of spouse along with appropriate education and nurture of the children. He based his assumption for future human enhancement on the findings of Pasteur regarding the enhancement of flowers by proper choice and cultivation. He also justified his argument by showing the studies of Galton in family trees of successful men. According to Makridis, eugenics was a branch of hygiene, which referred to groups (nations, races, humanity). It aimed at the conservation and multiplication of those organisms that had biological, physical and intellectual values. Eugenics was based on the principle that external factors and the environment were not the only factors of good health, but heredity played an equally important role in the evolution, progress and robustness of a race. However, he admitted that eugenic policies would not have obvious results sooner than their application to three or more generations.

As far as pregnant women were concerned, they should be aware of any information that would help them to protect themselves and their children. According to Savvas, the health of a newborn was threatened by its parents; syphilitic parents, for instance, could inhibit the development of the embryo or even cause stillbirth. After birth, most of the health problems were caused by malnutrition. In agreement with Savvas and in order to disseminate eugenics, A. Doxiadis encouraged activities which informed mothers, such as Mother's Day, Children's Week, giving awards for beautiful children, and financial aid for families having three or more children (Karakatsani and Theodorou 2010). Health and beauty contests for fitter families occurred in the USA during the first decades of the twentieth century. The contests also served the purpose of public education for child rearing and hygienic practices (Stern 2002; Bicchieri Boudreau 2005).

In this context, eugenics intersected with Adolphe Pinard's theory of 'puériculture' which was easily acceptable by the state, physicians and the public (Pinard 1908). Moisidis wrote a book on eugenics and puériculture using ancient Greek texts to validate his views (Moisidis 1925a). Puériculture offered the theoretical framework to form state policies for the protection of mothers and children. It included a programme of advice for prospective parents for the periods before conception, during pregnancy and after birth. State propaganda was based on advice on nutrition, care and hygiene of a pregnant woman and the newborn, which was, in fact, a popularisation of puériculture. Savvas insisted on the necessity to inform the illiterate about puériculture by simplifying and popularising it. Moreover, obstetricians and midwives should inform new parents about the protection of their child and teach young girls in schools about the necessity of breastfeeding and puériculture.

The first half of the twentieth century was characterised by a wide range of health problems in Greece. The most significant were infant mortality and the transmission of dangerous diseases. Due to limited funds, the Greek state was unable to react and deal with the problems quickly. Much effort was put into improving the effectiveness of the public health sector, but most of the legislation was never implemented. Physicians, paediatricians and scholars who were preoccupied with public health and hygiene often became excessive and proposed eugenic policies, like the prohibition of marriage to certain groups of people and state intervention in families. The bigger picture, though, shows efforts to confront the problems at their root and construct a regenerated Greek society consisting of healthy, strong and intelligent citizens. To this end, the objective was the protection of mothers and children, which became a priority and shaped public health policies.

Along with the collapsed public health system, the Civil War which followed the Second World War disturbed the social cohesion of Greece. The country was divided into two camps: the National Army and the Democratic Army of the Communists. The Civil War has had devastating consequences for the Greek nation, polarising society and thus resulting in hundreds of thousands of murders and excessive political violence from both sides (Close 1995). It ended with the Convention of Varkiza of 1949, which confirmed the defeat and disarmament of the Democratic Army. Peace was restored in 1950, but a large part of the society continued to be divided into communists and anti-communists. During the post-war period of social and political reconstruction, the ultimate goal was sociopolitical unity. Polarisation was the root cause of the national disaster of the Civil War which had to be avoided. Therefore, although the existence of anti-communism and anti-nationalism was undeniable, the majority of the Greek society and political authorities avoided public expressions of both in order to secure peace, unity and political stability. Modernisation and social transformation were the new tasks of the post-war Greek society. Despite the fact that the country avoided communism, the Civil War hindered its 'westernisation'. Greece had to reach the Western European countries. Thus, the need to recover from the wars and re-organise the state and the society was imperative. However, the most urgent undertaking was the health restoration of its citizens. Consecutive years of famine, epidemics and lack of hygiene resulted in a low health level. The mortality and morbidity rates reached a peak in the 1950s. The situation began to improve in 1951, when the Ministry of Social Care was re-organised. It was in this context that the Greek eugenics movement gained popularity.

NOTES

1. See a detailed analysis of the protection of motherhood in Greece during the first half of the century in: Karakatsani, Despina, and Vasiliki Theodorou. 2012. Eugenics, Childcare and Hygienic Concerns in Interwar Greece. In *Anthropological and Sociological Approaches of Health*, ed. Charalampos Economou and Manos Spyridakis, 483–510. Athens: I. Sideris.
2. More information on premarital medical certificate will follow in Chapter 5.

REFERENCES

Academy of Athens Presidents Since Its Establishment. http://www.academyofathens.gr/ecPage.asp?id=211&nt=18&lang=1. Accessed 4 Mar 2013.

Athens Medical Association Archive. 1952. *Proceedings of the Thirteenth Regular Meeting*.

Bicchieri Boudreau, Erica. 2005. 'Yea, I Have Goodly Heritage': Health Versus Heredity in the Fitter Family Contests, 1920–1928. *Journal of Family History* 30: 366–387.

Close, David H. 1995. *The Origins of the Greek Civil War*. New York: Longman.

Dardavesis, Theodoros. 2005. The School of Hygiene in Athens and Its Development into the National School of Public Health. *Iatrika Themata* 39: 24–31.

Dardavesis, Theodoros. 2008. The Historical Course of the Ministry of Health in Greece, 1833–1981. *Iatriko Vima* 115: 50–61.

Doxiadis, Apostolos, and Zoe Fragkou. n.d. *The Hygiene and Nurture of the Children*. Athens: n.p.

Economopoulos, Nikolaos. 1922. *Social Hygiene: Social Care-State Care*. Athens: Petrakis.

Economou, Charalampos, and Manos Spyridakis (eds.). 2012. *Anthropological and Sociological Approaches of Health*. Athens: I. Sideris.

Exarchopoulos, Nikolaos. 1931. The Diagnosis of the Level of Intelligence on the Basis of Experimental Research. A New Version of the Binet-Simon Test. *Academy of Athens Minutes*, 6. Athens: Academy of Athens.

Exarchopoulos, Nikolaos. 1950. *Introduction to Pedagogy*. Athens: n.p.

Gardikas, Katerina. 2008. Relief Work and Malaria in Greece, 1943–1947. *Journal of Contemporary History* 43 (3): 493–508.

Gardikas, Katerina. 2018. *Landscapes of Disease: Malaria in Modern Greece*. Budapest and New York: Central European University Press.

Hionidou, Violeta. 2006. *Famine and Death in Occupied Greece, 1941–1944*. Cambridge: Cambridge University Press.

Hirschon, R. 2003. *Crossing the Aegean: An Appraisal of the 1923 Compulsory Population Exchange Between Greece and Turkey*. Oxford: Berghahn Books.

Jensen, Arthur R. 2002. Galton's Legacy to Research on Intelligence. *Journal of Biosocial Sciences* 34: 145–172.

Karakatsani, Despina. 2011. Hygiene Imperatives: Child Welfare, School Hygiene and Puériculture in Greece (1911–1936). *Годишњак за друштвену историју* 3: 25.

Karakatsani, Despina, and Vaso Theodorou. 2010. *Hygiene Imperatives: Medical Observation and Social Care of the Child During the First Decades of 20th Century*. Athens: Dionikos.

Kiriopoulos, Ioannis. 2008. *Public Health and Social Policy: Eleftherios Venizelos and His Time*. Conference Proceedings. Athens: Papazisis.

Makridis, Nikolaos G. 1940. *For the Protection, Improvement and Ennobling of the Greek Race*. Athens: Anatoli.

Mazumdar, Pauline. 1992. *Eugenics, Human Genetics and Human Failings: The Eugenics Society, Its Sources and Its Critics in Britain*. London: Routledge.

Moisidis, Moisis. 1922. *Eugenics and Marriage*. Constantinople: n.p.

Moisidis, Moisis. 1925a. *Eugenics and Puériculture in Ancient Greece: Contribution to the History of Puériculture*. Athens: n.p.

Moisidis, Moisis. 1925b. *Woman: Hygiene of Marriage and Married Woman*. Alexandria: n.p.

Moisidis, Moisis. 1928. *Abortion in Ancient Greece: Forensic, Clinical and Pharmacological Study*. Athens: n.p.

Nikolaidis, George, and Spyros Sakellaropoulos. 2002. Social Policy in Greece in the Interwar Period: Events, Conflicts and Conceptual Transformations. *SAGE Open*. http://sgo.sagepub.com/content/2/4/2158244012461491. Accessed 22 July 2013.

Official Government Gazette. 1834. Royal Decree: About the Establishment of the Medical Council. 24.

Official Government Gazette. 1908a. Law 3252: About the Sale of Quinine. 60.

Official Government Gazette. 1908b. Royal Decree: About the Establishment of the Office of School Hygiene in the Ministry of Ecclesiastics and Preliminary Education. 239.

Official Government Gazette. 1912. Law 4029: About the Labour of Women and Minors. 46.

Official Government Gazette. 1914. Law 240: About the Administration of Education. 97.

Official Government Gazette. 1915. Law 346: About the Supervision of Public Health. 2.

Official Government Gazette. 1917. Law Enactment: About the Establishment of the Ministry of Social Care. 112.

Official Government Gazette. 1922. Law Enactment: About the Establishment of the Ministry of Hygiene, Healthcare and Perception. 269.

Official Government Gazette. 1926a. Law Enactment: About the Abolishment of the Ministries of National Economy, Hygiene, Healthcare and Perception. 11.

Official Government Gazette. 1926b. Law Enactment: About the Re-establishment of the Ministry of Hygiene, Healthcare and Perception. 286.

Official Government Gazette. 1929a. Law 4172: About the Establishment of the Independent Ministry of Hygiene. 201.

Official Government Gazette. 1929b. Law 4069: About the Establishment in Athens of the School of Hygiene and the Physicians' Postgraduate Studies on Hygiene Abroad. 94.

Official Government Gazette. 1932. Law 5733: About Social Insurance. 364.

Official Government Gazette. 1934. Law 6298: About Social Insurance. 346.

Official Government Gazette. 1937a. Law 965: About the Organisation of Public Health Institutions. 476.

Official Government Gazette. 1937b. Law 547: About the Reformation of Labour Laws. 98.

Papaioannou, Antonios Ch. 1939. *Student's Health Card*. Athens: n.p.

Pearson, Karl. 1909. *The Groundwork of Eugenics*. London: Dulau & Co.

Pearson, Karl. 1912. *The Problem of Practical Eugenics*. London: Dulau & Co.

Pinard, Adolphe. 1908. *De la Puériculture*. Lyon: Imprimeries Reunies.

Sarantis, Konstantinos. 2009. Ideology and Political Character of the Metaxas' Regime. In *Metaxas and His Time*, ed. Thanos Veremis, 58. Athens: Evrasia.

Savvas, Konstantinos. 1928. *Handbook of Hygiene*. Athens: n.p.

Stern, Alexandra Minna. 2002. Beauty Is Not Always Better: Perfect Babies and the Tyranny of Pediatric Norms. *Patterns of Prejudice* 36 (1): 68–78.

Stoyiannidis, Yiannis. 2016. The Social Issue of Tuberculosis and the History of Sanatoria in Athens, 1890–1940. PhD thesis. Volos: University of Thessaly.

Theodorou, V. 2002. Doctors Against the Social Issue: The Anti-tuberculosis Struggle in the Beginning of the Twentieth Century (1901–1926). *Mnimon* 24: 145–178.

Trubeta, Sevasti. 2013. *Physical Anthropology, Race and Eugenics in Greece (1880–1970s)*. Leiden: Brill.

Weindling, Paul. 1997. Philanthropy and World Health: The Rockefeller Foundation and the League of Nations Health Organisation. *Minerva* 35: 269–281.

CHAPTER 3

The Hellenic Eugenics Society

As a consequence of the consecutive wars from the beginning of the twentieth century until 1949, living conditions continued to be very difficult in the early 1950s, and the general health of the population was at a historically low level. What is often ignored by historians is the effect of the famine from May 1941 to April 1943, which not only caused numerous deaths, but also sterility (Hionidou 2006). It has been argued that the chronic malnutrition during the two years of starvation affected the male population more than the female and children (Valaoras 1946).

Furthermore, the public healthcare system was disorganised and poor (Burnova 2005). During the period 1940–1951, the Ministry of National Hygiene and Perception was renamed and split into different sectors and reunited several times. After a short period of stability, during 1951–1964, it followed the same course of continuous changes of name and ministers. It is remarkable that during the period 1917–1982, 102 ministers of health were appointed by the state to manage the vulnerable portfolio of public health and hygiene (Dardavesis 2008).

The need to promote health and hygiene guidelines to the public became evident when the Law 2032/1952 which provided for the creation of a new Public Education Service, belonging to the then Ministry of Welfare, responsible for public education and propaganda for health and hygiene was passed in 1952. The purpose of this service was to undertake a campaign to address the problems of personal and public hygiene, prevention of diseases and maintenance of physical and mental

© The Author(s) 2019   45
A. Barmpouti, *Post-War Eugenics, Reproductive Choices and Population Policies in Greece, 1950s–1980s*,
https://doi.org/10.1007/978-3-030-03568-6_3

health. The service was also willing to cooperate with any public or private initiative towards fulfilling its aims.

The Athens Medical Association (hereafter AMA) and other non-governmental institutions took the opportunity to bring problems of hygiene to public attention by organising lectures and conferences. Among these was the union of several non-governmental associations, women's clubs and scientific societies, which was given the provocative title: 'National Crusade of the Scientific and Social Organisations for the Psychological, Mental and Physical Health of Greek People' (*Εθνική Σταυροφορία Επιστημονικών και Κοινωνικών Οργανώσεων δια την Ψυχικήν, Πνευματικήν και Σωματικήν Υγείαν του Ελληνικού Λαού*). This non-official movement was founded by sixteen independent, non-governmental associations and organised a series of thirty-five lectures from 26 May until 29 June 1952 (Deltion Iatrikou Syllogou Athinon 1952b). It included lectures on the role of Greek women in society and the family, premarital health certificates, directives for mental health, alcoholism, drug addiction, neurotic children and sex education. Speakers included well-known eugenicists and future members of the Hellenic Eugenics Society (HES) such as Popi Spelioti-Bazina, a renowned feminist, gynaecologist (Spilioti-Bazina 1952); Moisis Moisidis, a physician; Konstantinos Konstantinidis, a physician, Konstantinos Katsaras, a psychiatrist and Nikolaos Drakoulidis, also a psychiatrist.

In their attempt to disseminate the rules of hygiene and preventive medicine, the creation of a Greek Eugenics Society was an idea initially conceived and developed by the AMA. The AMA was the largest of its kind in Greece, both in terms of the number of its members and in the scope of its activities. Discussions about the improvement of personal and public hygiene as well as preventive medicine were abundant throughout the twentieth century. The association tried to improve the health of the Greek population, particularly the poor. One of its targets was to familiarise the public with modern ideas of health and hygiene. Following the model set by the World Health Organisation, the AMA celebrated the health day, having a different topic every year. The World Health Organisation celebrates the World Health Day on 7 April every year in remembrance of its first Assembly and founding in 1948 (WHO 1949). In 1953, the topic of Health Day was sanitation. A year earlier, in 1952, the AMA organised a competition among physicians for the best-written non-professional pamphlet on personal hygiene showing their interest in passing the principles of hygiene to the general public (Deltion Iatrikou Syllogou Athinon 1952a).

Surprisingly, none of the official minutes of the AMA included any notion or remark about the creation of a eugenics society or anything about eugenics in general. The official minutes of the meetings of the Board of Directors of the AMA dealt with internal affairs, inspection of physicians' practice and financial matters.[1] Nor did the bulletin of the AMA refer to the foundation and activities of the HES. However, it published articles by Konstantinos Gardikas (1952), a long-standing eugenicist, and also inspired new converts to eugenics, such as Vasilios Valaoras (1953a, b) and Spyros Doxiadis (1953). Notwithstanding the absence of reference to the HES in the AMA publications and official documents, Athanasios Mantellos, President of the AMA from 1951 to 1953, set up the foundation of the HES under the auspices of the AMA. The first meetings aimed at the foundation of the eugenics society took place at the premises of the AMA in Athens. As Mantellos claimed during the meeting on 29 March 1953,[2] the HES was going to be part of AMA's work towards the protection and pursuit of the prosperity of the Greek nation. Obviously, the creation of the eugenics society was a natural outcome of this growing interest in Greece in hygiene and population problems. Louros' personal archive includes valuable information about the first meetings of physicians at the premises of AMA. The archival material allows us to dig deep in the physicians' eugenics mentality; to understand the reasons behind their desire to form a eugenics society at this time period and their ideas on the ways to implement their plans.

1   AIMS AND SCOPE

The first documented meeting, whose purpose was to discuss the possibility of establishing a eugenics society in Athens, was held on 29 March 1953. A few more meetings followed in 1953 but complete minutes are available for only the first two, held in March and May that year. Although the creation of a eugenics society was Mantellos' idea, it was Maro Kanavarioti, the would-be first secretary of the HES, who put great effort to realise it. Kanavarioti remains an enigmatic figure in the history of post-war Greek eugenics. We know little about her life and activities. As the first secretary of the HES, however, she handled its international and domestic relationships, but at the same time, the statutes of the HES refer to her as a 'housewife'. However, it is assumed that she studied medicine but maybe she did not practice it. There was an indication that she was a physician in a letter from Evangelos Danopoulos,

Professor of Pathology at the University of Athens, where he addressed her as 'colleague'.[3] Alexandros Stavropoulos, who wrote a book on the first premarital consulting centre in Greece, also suggested that Kanavarioti was a physician (Stavropoulos 1970). Moreover, she wrote in excellent English, which was rather unusual for women in Greece during the 1950s. According to the *Eugenics Review*, she was also a fellow of the Eugenics Society in Britain (Editor 1954). Supposedly, she came from a wealthy family or spent some time abroad, maybe in the USA, where her daughter lived, or possibly in the UK. Lelia K. Washburn, Kanavarioti's daughter, moved to New York in the late 1940s. She received her Master's degree in American Studies from Harvard University in 1953 and became a Professor of Ancient and Modern History at the American University, Washington, DC (McDonough 2013).

On 22 May 1953, Kanavarioti sent a personal letter to the American demographer Dr. Pascal K. Whelpton, who at that time was Director at the Population Division of the United Nations (Durand 1964). From this letter, we know that Whelpton had visited Greece in December 1952 and gave a lecture on issues of population and eugenics in Athens. Kanavarioti informed Whelpton about the positive outcomes of his lecture in the King George Hotel in Athens. The Greek physicians were interested in promoting public awareness of the growing population problems of Greece.[4] As already mentioned, it was Whelpton's visit that triggered the discussion among a group of scientists who gathered in the premises of the Medical Association in Athens about the creation of a eugenics society on 29 March 1953. Less than two months later, there was a unanimous decision of forty scientists to create a eugenics society. Eugenics was then regarded by the Greek physicians as an effective solution to the population problems of Greece, such as the low birth rate and the exceeding number of induced abortions.

Interestingly, Kanavarioti referred to eugenics as a new topic in Greece. Contrary to what she seemed to believe, eugenics was not new to Greece. As already discussed, during the interwar period, Greek physicians and anthropologists adopted eugenic ideas and debated various eugenic programmes for Greece. Yet, there was no Eugenics Society in Greece prior to the Second World War. This probably was why Kanavarioti described eugenics as a 'new topic' in Greece. Although eugenics gained supporters during the interwar period, the decision to institutionalise it was taken much later. The contribution of foreign experts was crucial to the decision to form an official Eugenics Society aiming at playing a consultative role to the Greek state's population policies.

As revealed by the minutes of the first meeting, the founding members were aware of the existing opposition to eugenics, a fact also mentioned by Kanavarioti in her letter. Without making it clear, she wrote to Whelpton that the members of the society anticipated opposition from some groups in Greece. It is, however, unclear whether she had a specific scholarly hostility in mind or perhaps she assumed that the general public would react negatively to the Greek eugenic movement. The reluctance to accept the creation of a Eugenics Society in Greece seems justified, not least because of the very recent memory of the eugenic policies of Nazi Germany and the Holocaust of the Greek Jews. Moreover, the political and social segregation caused by the Civil War was to be avoided during the post-war period. Given that the Civil War ended in 1949, at the time of the foundation of the HES in 1953 internal peace was still fragile. The establishment of a Eugenics Society in Greece was expected to generate negative reactions, either due to its name, 'eugenics', which took such a negative meaning after the Second World War, or due to the fear that it might support policies deemed to be against political cohesion and in favour of social discrimination.

It is due to Kanavarioti's letter to Whelpton that we now have details about the formation of the HES. It began with a 'provisional committee' whose role was to draft 'the society's charter, in which population problems would be clearly stated and included'. The committee reported on its activities at a subsequent meeting two weeks later, when the intention to establish a Eugenics Society was reaffirmed. Indeed, less than a year after Kanavarioti sent Whelpton the letter, the HES was officially established. As Kanavarioti noted, 'this [was] the beginning of a new era in this respect for this part of the world'.[5]

Clearly, Whelpton's visit had a particular purpose, namely to raise awareness of the importance of eugenics and population research in Greece. Although the reason why Whelpton chose to motivate the Greeks to establish a Eugenics Society was unclear, it was certainly not a coincidence that few months prior to his visit in Athens the regional department of the International Planned Parenthood Federation (IPPF) was established in London. To this effect, Kanavarioti thanked Whelpton for his interest in helping organising the eugenics movement in Greece, assuring him that she will continue to inform him about the progress of their activities. She also let him know that his future contribution would be welcomed by the members of the preliminary committee. Ultimately, the creation and the subsequent activity of the HES clearly demonstrated that Whelpton had fulfilled his aims.

From the style and nature of their correspondence, it seemed that Kanavarioti knew Whelpton already. Kanavarioti eventually became the contact person between Whelpton and the Greek eugenicists. In addition to her contact with Whelpton, there is a handwritten letter sent by Kanavarioti to William Vogt, dated 10 March 1953, referring to her visit to Stockholm, probably to attend the meeting of the IPPF showing the broad spectrum of her contacts. The letter confirms that Kanavarioti was already familiar with foreign institutions regarding birth control and eugenics.[6]

Even from the initial discussions of the first group of eugenicists, it became clear that the society would aim at promoting birth control and eugenics. Although there was consensus about the need to introduce eugenic policies in Greece, disagreement persisted over which ones were necessary. During the 1950s, the population trends were not clearly indicated in statistical figures because the country experienced a significant drop in mortality rates, which was translated into population increase. At the same time, the birth rate was gradually decreasing. However, this was not evident because birth decline was outweighed by the increase in life expectancy. Therefore, although the overall quantity of the population seemed to increase, the birth rate was low. In addition, there was a wave of internal immigration from the countryside to the city centres, resulting in overpopulated cities, particularly Athens. As a result, different, often opposite, opinions were voiced during the first meetings of the HES. Eleftheriadis, a physician, was against the control of reproduction, as it would be against the interest of the nation, because Greece had already experienced low birth rates. Instead, he endorsed quantitative reproduction, by which he meant numerous births, rather than fewer and better cared for children.

Valaoras, the famous physician and biostatistician, on the other hand, believed that since mortality rates had decreased, some policies regarding birth control should be adopted by the state. He thus answered Eleftheriadis' claim that birth control would lead to low birth rates and affect population growth in Greece. Moisidis also endorsed birth control and insisted in founding a society responsible for dealing with issues of procreation and eugenics. As mentioned in the previous chapter, Moisidis was already a famous eugenicist, having published a number of articles and books on eugenics since the beginning of the twentieth century (Moisidis 1925a, b, 1928). Moreover, he was aware of the function of similar societies abroad and desired the same for Greece.

Mantellos interpreted eugenics within the framework of state-supported policies aimed at encouraging the birth of healthy children. Furthermore, he identified eugenics not only with the birth of healthy children, but also with the ideal living conditions for raising a child. As the President of the AMA and a physician himself, Mantellos wanted to enlist the new Eugenics Society's help for the AMA's efforts towards the improvement of the living standards and the health of the Greek people. He thus argued that the Eugenics Society should not limit its activities to birth control propaganda, but be active in many other areas of public health as well.

Equally important, all participants agreed that they wanted to establish a good relationship with the state. Maslarinou, a female physician, was the first to mention a possible negative attitude by the state towards the newly founded Eugenics Society. As noted earlier, the same view was expressed by Kanavarioti in her letter to Whelpton. Drakoulidis, a psychiatrist, mentioned that the Greek state had not implemented any effective policies to tackle the population problem of the country, therefore it needed to be better informed. The HES could play exactly this role, namely to advise the state in these matters. With regard to political intervention, Drakoulidis recalled an incident that had happened 30 years previously (ca. 1923), when he had delivered a speech about the campaign against venereal diseases. While he was talking, a policeman interrupted him just because he had the power to do so. Drakoulidis shared with the group another incident indicating the state's power over the academia. In 1932, he attended a conference organised by the Commercial Chamber, when the former Prime Minister Eleftherios Venizelos expressed the view that the state should punish infected people rather than help them. Obviously, as a physician, Drakoulidis disagreed with Venizelos' opinion. He used these incidents to argue that if the state authorities did not agree with the new Eugenics Society, they would not be deterred from restricting its establishment and activity. It would be better, Drakoulidis suggested, to be on good terms with the Greek state. The matter of the relationship with the state and the church would be repeated many times during their following meetings.

In addition, Fylaktopoulos, a psychologist, who agreed with Drakoulidis, added that they should also establish good relations with the Orthodox Church. The Holy Synod of the Greek Orthodox Church was already informed about the HES's activities. Although constitutionally not pervasive or authoritative, the Orthodox Church played an important role in the Greek people's lives. As the dominant religion in Greece, Orthodoxy was influential over daily affairs. According to Fylaktopoulos, the HES should be very well organised before getting in touch with the state and the church in order to decide which would be the optimal 'form' of the eugenics movement. He estimated that up to two years were required for this purpose. Fylaktopoulos also raised the issue of 'national duty'. The members of the HES perceived their activity as their duty to protect the nation. National protectionism has always been part of the eugenics rhetoric even from the early twentieth century. According to Quine, 'men of science and medicine saw themselves as the guardians of the future with a mission to apply their knowledge socially for the common good' (Quine 1996). Indeed, during this historical period in Greece, the circumstances were favourable for such attitudes. On the other hand, in the eyes of a largely uneducated society simultaneously facing health deterioration, the well-educated, respectful and often wealthy physicians represented their hope for national regeneration.

Therefore, committed to their mission, the members of the HES decided that a second, more formal meeting was necessary, in order to begin with the actual organisation of the Eugenics Society. The themes proposed for discussion were overpopulation, demographic problems and 'conscious' reproduction echoing the international trend on the preoccupation with population problems. To this end, four members, Mantellos, Moisidis, Fylaktopoulos and Kanavarioti, formed a temporary committee to undertake the preparation for the second meeting, held on 19 May 1953. Forty-six people attended to discuss and decide on the foundation of a Eugenics Society. Most of them were physicians and among them there were four women. The official statutes of the new society, however, were signed only by twenty-seven of them.

The second meeting was important for two reasons. Firstly, the first Executive Board was formed, with Mantellos as president, Kanavarioti as secretary, and a temporary committee of seven members, including Spyros Doxiadis, Konstantinos Katsaras, Konstantinos Konstantinidis, Andreas Pournaras and Konstantinos Saroglou—all physicians. Secondly, on this occasion, Mantellos announced that it was Kanavarioti's idea to establish an organisation for the study of birth problems and population movement in Greece from the scientific, familial, social, financial and national point of view. Members of the AMA and other scientists were impressed by this idea and agreed that such a society would play a vital social and national role in the study and evaluation of findings regarding eugenics and the biological progress of the Greek nation.

On behalf of the AMA, Mantellos repeated that the attempt to establish a Eugenics Society in Athens related to its social work; thus, the AMA would offer its premises to house the new society. Furthermore, Mantellos emphasised the necessity of taking specifically oriented actions towards the biological enhancement and the improvement of the living conditions of the Greek nation adapted, however, to the current socio-economic conditions.

Among those who agreed with the establishment of a Eugenics Society was Nikolaos Tsampoulas, who proposed the cooperation with similar organisations and the state; the idea of giving the prospective Eugenics Society the role of a scientific committee intended to advise the government in matters of population eugenics was also shared by Pournaropoulos and Antonopoulos. Konstantinos Katsaras argued that eugenics was a very important issue, particularly for the poor. He added that a eugenics association could aid the Greek state to implement its policies aiming at the 'cure of great social injuries'. In addition, Mrs. Chrysoula Ioakimidou claimed that the birth of healthy children was of ultimate importance to the nation. Moreover, Valaoras argued that there were already many governmental and non-governmental associations dealing with the health of pregnant women, mothers and children, although inadequately. If finally established, the Eugenics Society should more intensely pursue the enlightenment and education of both those intending to marry and the newly married, because it was at this point that hereditary, biological and environmental factors should be considered and evaluated according to the quality and quantity of the population. Georgios Gonos envisioned a Eugenics Society which would examine theories and practices of eugenics and adapt them to the Greek context, aimed at the biological and social prosperity of the Greek people. Adding to Gonos' perspective, Evangelos Danopoulos mentioned that the new society should assess both positive and negative eugenics and propose viable solutions to demographic problems. Above all, it would be a scientific society intending to educate the public. Another issue raised by Danopoulos was the imitation of foreign examples, namely the work of other European Eugenics Societies. At the time, the British and the American Eugenics Societies were still active and supposedly Danopoulos had these two in mind. The collaborations that followed among these two and the HES verify this assumption. Georgios Igoumenakis argued that it was all doctors' duty to deal with problems of eugenics. Telling, the purpose of the new society would be to advise the state about degenerative factors of the population, such as venereal diseases, in order to eradicate them accordingly. Moreover, Ilias Katsaniotis claimed that the country had already suffered from demographic problems, so the problem of eugenics should be profoundly examined. Spyros Doxiadis underlined the advantages of precise public education by the use of statistics showing the influence from Galtonian eugenics. He added that problems like urbanism and child mortality should be considered as well.

It is also interesting to discuss Dionysios Travlos' views. Travlos, Professor of Gynaecology at the University of Athens, argued that the Eugenics Society should pursue achievable goals, meaning that their plans should be adapted to the Greek lifestyle and living standards. He pointed out that eugenicists should opt for a gradual change and not an immediate one, simultaneously leaving out unrealistic theories. He was the only one who focused on the practical aspects regarding the Eugenics Society's potential list of activities and proposed a reasonable plan of action from simple to the more difficult tasks.

Mantellos then summarised the opinions of the participants and came to the following conclusions: the Eugenics Society should be primarily an advisory board, a scientific association, but its ultimate purpose would be to lobby for the implementation of its findings in specific legislative, administrative and social policies. Therefore, its members should be not limited to physicians, but also sociologists, economists, journalists and mothers. The society thus formed would be named HES (*Ελληνική Εταιρεία Ευγονικής*).

A group of members, including Kanavarioti, Mantellos, Travlos, Fylaktopoulos, Doxiadis, Saroglou, Konstantinidis, Katsaras, Pournaras, Tsampoulas and Moisidis, were asked to prepare a draft of the statutes for the next meeting. Eventually, the statutes had fourteen articles and were deposited in the Court of First Instance for legal approval. During the meeting, the aims of the society were also outlined, including: (a) the research and study of problems of eugenics in Greece; (b) the dissemination of eugenics; and (c) the cooperation with the state and non-governmental organisations regarding public education on matters of eugenics. These aims were also outlined in the first article of the official statutes.

2   THE STATUTES

Louros' personal archive included the original statute as it was deposited in the Court of First Instance in Athens. It was signed on 16 July 1953 and officially recognised by the Court on 19 April 1954, having the register number 7479/19/4/54. The text of the statute was published alone as a booklet in 1956, three years after the signing of its final version by the members. The importance of the study of the statutes is attributed to the fact that actually summarised the eugenic views of the time. Given that it was written and agreed by the elite of physicians and scholars interested in eugenics, it is reasonable to argue that it represents the dominant views of post-war Greek eugenics. The structure and function of the HES were based on the statutes. The final version of the statutes was signed by the following twenty-seven members, namely: Alivizatos Gerasimos, Antonopoulos Dimitrios, Valaoras Vasilios, Danopoulos Evangelos, Spyros Doxiadis, Igoumenakis Georgios, Kaminopetros Ioannis, Kanavarioti Maro, Katsaras Konstantinos, Katakouzinos Evangelos, Konstantinidis Konstantinos, Malikiosis Xenofon, Mantellos Athanasios, Moutoussis Konstantinos, Moisidis Moisis, Panayiotou Panayiotis, Papadakis Antonios, Pournaras Andreas, Pournaropoulos Georgios, Saligkarou Pasithea, Saroglou Konstantinos, Spiliotis Panagiotis, Stefanou Dimitrios, Travlos Dionysios, Tsampoulas Nikolaos, Fylaktopoulos Georgios and Choremis Konstantinos. Even though most of them were famous physicians, two of them, Doxiadis and Choremis, were the most renowned paediatricians in Greece. Furthermore, Kaminopetros and Moisidis were two of the protagonists of the pre-war eugenics movement in Greece. Interestingly enough, Louros was not among these members now did he participate in the first meetings. Last but not least, only two of the founding members were women, Kanavarioti and Saligkarou.

The officially approved statutes of the HES consisted of fourteen articles. Article one referred to the title, base and purpose of the HES. As already mentioned, the official name was 'Hellenic Eugenics Society', based in Athens. The aims were summarised in three main categories:

1. The study of issues of eugenics in Greece and their connection to the quality and quantity of the population, on the basis of its genetic factors and the specific environment of the country.
2. The communication of the acquired knowledge from these studies to the government in order to implement national policies

regarding these matters, intended to promote good psychosomatic qualities in the Greek population.

3. In cooperation with the state and other social organisations, to promote public education to avert possible degenerating factors, if and when developed; to advance the harmonious growth of the Greek population within the economic and social potentialities of the country and, finally, to improve the living standards of the Greek family in general.

Obviously, the HES did not aim at becoming an association limited to exchanging information among scientists but disseminating their knowledge to the general public. Eugenics' aim for public education was firstly set out by Sir Francis Galton; thus, it was incorporated in every Eugenics Society's plans. The importance of public education was also evident in the first name of the British Eugenics Society, which was 'The Eugenics Education Society' (Galton 1909; Wright Gillham 2001; Turda 2007). The aims of the HES also echoed Leonard Darwin's suggestions for a successful Eugenics Society already set out in 1921:

> The main aim of eugenical societies should be [...] to formulate a sound eugenic policy based on existing genetic knowledge, and then to promote the translation of every advance in eugenic theory into general practice. (Darwin 1921)

A close inspection of the aims of other European Eugenics Societies verifies their aspiration to spread the word of eugenics to the general public (Schneider 1990; Turda 2014); thus the HES followed their example. In order to achieve their goals, the members of the HES used every means possible, including meetings and conferences, publications, radio broadcasts and educational films.

Articles two to five referred to membership. Members were divided into: honorary, regular and corresponding. Honorary members were people who would significantly contribute to the dissemination of eugenics or aid the work of the HES. They would have the same rights as the regulars, but without the right of voting. They had to be voted in by at least ten regular members. Regular members had to be voted in by at least two members; accept that statute; and pay their subscription. Corresponding members were individuals who lived outside Athens and were voted in by two regular members. If they ever moved to the capital, they received the same rights as the regular members.

Article six referred to the General Assembly of the HES, which would be responsible for every aspect of the HES's work. The members would be informed about the General Assembly by written invitation or by a publication in a daily newspaper in Athens at least eight days in advance. During each meeting, the voting would be open. It was obligatory that the General Assembly would be arranged every January to discuss financial and other reports about the activities of the HES.

Articles seven to eleven referred to the administration. Apart from the president, who represented the HES on every occasion, there was an Executive Board, which consisted of the vice-president, the secretary, the treasurer and seven members. Moreover, the specific duties of the president, the secretary and the treasurer were defined in Articles eight to ten.

Article twelve referred to revenues. These were: subscriptions and dues of the regular members, as well as their exceptional dues, donations, savings from publications of the HES and any other income.

Article thirteen referred to general terms, such as that the statutes were passed by the founding committee of the HES. Every aspect that was not included in the statute would be undertaken by the General Assembly. The closing of the HES would occur only if a three-quarters majority of its members decided it and its belongings would be transferred to the Academy of Athens and the final article fourteen declared that this statute was approved by the General Assembly on 16 July 1953.

The text of the statutes covered every important aspect regarding the activity and functions of the HES and was deposited in the Court of First Instance by the HES' lawyer, Nikolaos Stampolitis. In the light of the solemnity of the statutes, it was obvious that the HES was a union of eminent scientists, sharing the desire to improve the quality and quantity of the Greek people. They did not want to act independently, but in accordance with the Greek legal framework and with the state's approval. During the period of activity of the HES, the members made every effort to secure a good relationship with state authorities. At times, however, acting according to the state's plans undermined the realisation of their own.

3   THE INTERNAL STRUCTURE

On 11 December 1953, a common letter was sent to the members of the HES, signed by both Mantellos and Kanavarioti. It provided information regarding the prospective activities of the HES and the preparation of its next steps. According to the contents of that letter, the HES had already managed to form a plan of action, contact similar societies abroad and deposit the statutes in the Court of First Instance for approval. The letter was accompanied by a list of subjects that the HES would focus on, which were agreed by the temporary Executive Board during several meetings. The members were asked to examine the list and propose their possible contribution in relation to any of them, no later than the end of the year. In this way, the Executive Board would be able to make a schedule of conferences and meetings in the following year.

The enclosed list with subjects of research areas was divided into four categories. The first subject was genetics, which included hereditary diseases, intelligence tests and the premarital health certificate. The second one was the environmental influence on human development. The third one was about population problems, such as global and local population's movement tendencies. The fourth category was about economy and lifestyle, such as production, nutrition, residence, education, entertainment, intellectual productivity and the cost of living. Evidently, they had in mind a wide spectrum of activities as was later depicted in the choices of topics of discussion during their public conferences.[7]

In addition, a General Assembly was to be held the following year in order to examine the members' reactions to these matters and elect a tactical Executive Board. The outcomes of the discussion of the aforementioned topics would be destined to the government, the press and propaganda. Judging from this letter, the HES did not take any serious actions before its official approval on 19 April 1954. According to the invitation sent on 6 March 1954, the General Assembly meeting held on 22 March 1954. They announced its temporary Executive Board and the regular committee. During the General Assembly, Konstantinos Saroglou delivered a speech about the aims of eugenics and the plans of the HES. Another invitation dated 31 June 1954 suggests that the General Assembly did not elect the president, vice-president, secretary and treasurer of the Executive Board on 22 March. This was the purpose of a new assembly, which was organised on 6 August 1954. The invitation was signed by Mantellos and Kanavarioti.

The HES' leading body was the General Assembly; major decisions about the function and activity of the HES were taken only by the General Assembly. The Executive Board, on the other hand, was the directorial body of the HES; it consisted of the president, the vice-president, the General Secretary, the treasurer and seven members.

The first Executive Board (1954–1957) included eminent academics such as its President Nikolaos Louros, Professor of Obstetrics-Gynaecology;

Vice-President Georgios Pantazis, Professor of Zoology; Treasurer Spyros Doxiadis, Professor of Paediatrics, Konstantinos Katsaras, a psychiatrist; Konstantinos Konstantinidis, Professor of Psychiatry and Neurology; Athanasios Mantellos, a physician and former President of the AMA; Panayiotis Panayiotou, Professor of Obstetrics-Gynaecology; Konstantinos Saroglou, Medical Director of the PIKPA; Georgios Fylaktopoulos, Professor at Athens College; Konstantinos Choremis, Professor of Paediatrics, with Maro Kanavarioti acting as secretary. All of them played a crucial role in the dissemination of eugenics in post-war Greece, when eugenics was no longer attached to physical anthropology but to other disciplines, such as gynaecology and paediatrics. The Executive Board directed the activities of HES on all levels; namely the organisation of the meetings, the sending of invitations, the contact with domestic and foreign organisations and institutions and many more duties. The Executive Board prepared the topics of discussion to put forward to the General Assembly, including the annual budget. The composition of the Executive Board changed every two to three years; however, only some of the members were replaced, not its entire membership. During the period from the first Board 1954–1957 until the Board of 1965–1967, there is a nucleus of members who maintain their position for many years. For example, Louros remains the president and Pantazis the vice-president; Doxiadis was the treasurer from 1954 until 1965, the same period when Panayiotou was a member of the Board. Konstantinidis and Fylaktopoulos remained members for the whole decade, and Adamopoulos entered the scheme in 1957 and remained until 1967 (Hellenic Eugenics Society 1977).

By the time of the election of the first Executive Board, a new era began in the history of the HES, primarily due to the prestigious figure of Louros and Kanavarioti's impressive work. As far as other meetings are concerned, there is only indirect information taken either by letters and notes of the participants or from the official statutes. By 1955, there had been two crucial changes; firstly Louros succeeded Mantellos as president; and secondly the house of the HES was transferred to Alexandra Maternity Hospital. In fact, the HES was then totally disassociated from the AMA. Mantellos was President of both the AMA and the HES until 1954, when he was appointed General Director of the Ministry of Social Care and abandoned both posts. However, he remained a member of the Executive Board of the HES.

The statutes were officially approved by the Greek state on 19 April 1954, but the new Eugenics Society was only announced to the general public at the beginning of 1955, when a letter was sent to the popular daily newspaper *Ta Nea* to announce its founding.[8] The notice was signed by Nikolaos Louros, the new president, and Kanavarioti, the secretary. Kanavarioti remained in this post until 1959, when Marios Raphael succeeded her. Another publication in the daily press was an announcement signed by the lawyer representing the HES, Nikolaos Stampolitis, in the newspaper *Apogeymatini* in March 1955 (Stampolitis 1955).

While the first meetings, which aimed at the foundation of a Eugenics Society in Greece, took place at the beginning of 1953, its official establishment came a year later. In April 1954, its statutes were approved by the Greek state and then it became more active. However, the pivotal point was the elections of the Executive Board in August 1954, when a new period followed, under the leadership of Louros. Due to the fact that the proceedings of the first meetings of the HES are not published yet and the fact that the HES was largely identified with Louros in the following years, it is mistakenly known that he was the first president. As evidence shows, however, Mantellos was the first president, Louros the second, Doxiadis the third but only for a year and last was Ioannis Danezis.

During the first years of Louros' presidency, the HES gained popularity and gradually became known in Greece and abroad. However, the minutes of the gathering of the Executive Board in January 1958 revealed the uneasy situation of the HES during the period 1957–1958.[9] First of all, Louros announced Kanavarioti's succession by Marios Raphael. This marked a transitional period, when the Executive Board had to be reorganised after its first synthesis during the period 1954–1957. Kanavarioti was the key person during the first three years of the HES, but she resigned and left for the USA, probably due to familial reasons. Louros took over handling both domestic and international affairs. He expressed to the rest of the members his disappointment about the little progress the HES had made during 1957 and the indifference of members in dealing with eugenics. At this point, it was as if he was alone in the effort to disseminate eugenics, but soon the HES made significant progress with the organisation of successful conferences the following years.

At the time, Louros expressed his cautiousness about the future of the HES, while Konstantinidis, Saroglou, Goutos, Adamopoulos and Fylaktopoulos shared the view that the issue of the dissemination of eugenics was delicate and often met with disapproval. Therefore, it was not a coincidence that many of the members were unresponsive towards the

HES' activities. However, they unanimously decided that they would continue their work as other similar societies had already done. In order to alter the difficult situation, they resolved to meet more often; to increase funding; and to attract audiences by inviting Joseph van Vleck, who was a member of the Governing Body of the IPPF to give a lecture in Athens.[10]

In his effort to raise awareness on eugenics, Louros announced his idea of forming a Working Committee, a subgroup to deal with public engagement and contact with lay people and institutions. The new committee's responsibility was to maximise the impact of the HES to the wider public. They had to report their plans and progress to the Executive Board and request approval for further actions. The first members of this committee were the physicians Dionysios Kaskarelis, Olga Chrysostomidou, Dimitrios Papaloukas, and the sociologists Artemis Emmanouel and Marios Raphael.

During the first meeting of the Working Committee in 1958, the members decided to work upon specific issues of eugenics, which allegedly appealed to the general public.[11] Given that agreement with the government was mandatory, they planned to urge the government to adopt precise and long-term population policies, fitting the social, religious and economic situation in Greece. The HES would then act according to this official population policy, avoiding a deviation from the government's position. The Working Committee endorsed (negative) eugenic ideas, such as the avoidance of procreation in cases of disease or special conditions under which procreation would be harmful both for the parents and the child because the biological improvement of the new generation was an imperative for the members of the HES. Moreover, the new committee would undertake the education of different social strata in urban centres and in the countryside directly at schools and workplaces because public education was one of the main targets of the HES. The members of the committee decided that propaganda should be divided into three separate categories, each reaching a different target group. Among them, the incorporation of eugenics to the health professionals' education was the cornerstone. Public discussions and conferences came next on the list and finally the use of mass media, such as radio, newspapers, leaflets and films to reach even the least educated people. The above-mentioned ideas were in fact a reiteration of the classic eugenic arguments and similar to the aims stated in the statutes of the HES.

The second meeting of the Working Committee was held on 29 January 1958 and included the idea of cooperating with scientific societies, such as the Medical Society (*Ιατρική Εταιρεία*), the Obstetrics and Gynaecology Society (*Εταιρεία Μαιευτικής και Γυναικολογίας*) and the Paediatric Society (*Παιδιατρική Εταιρεία*), and aiming to give lectures on eugenics during these societies' gatherings. The Working Committee made a list of possible lecturers for the academic audience; the most suitable for the purpose were the gynaecologists and paediatricians of the HES, namely Panayiotou, Triantafyllopoulos, Antonopoulos, Danopoulos, Doxiadis, Konstantinidis, Moutousis, Saroglou, Travlos, Vlissidis, Malamos, Katiforis and Kaskarelis.

The most difficult task, however, was organising the lectures intended for a non-academic audience. Suitable places for this purpose were schools, workplaces, factories, municipality buildings and regional health/well-being institutions. Marios Raphael undertook the responsibility to contact these facilities and arrange the lectures. Among others, the suggested topics of discussion included the anatomy and physiology of the reproductive system, premarital hygiene of men and women, the prerequisites for allowing or prohibiting marriage, the special conditions under which procreation should not be encouraged, advice on the hygiene of pregnancy and the hygiene of newborns and children.

Within a week, the Working Committee met to discuss its progress on 5 February 1958.[12] The central person of the third meeting was Dionysios Kaskarelis. He informed the rest that Louros agreed to include subjects of eugenics in his academic lectures at the Medical School of the University of Athens. Moreover, he was going to contact associations similar to the HES and it was his idea to put short, recorded propaganda messages in waiting rooms of health institutes. Louros, already in his sixties, was a prestigious figure at the Medical School and in the Athenian society more generally. At the same time worked as a Scientific Director and Obstetrician-Gynaecologist at the Alexandra maternity hospital. The members of the Working Committee unanimously decided that two subjects would be more fruitful to non-academic audiences: (a) the meaning of 'good quality' in procreation, which would include aspects of anatomy, physiology, good and bad conditions for procreation and hygiene of pregnancy, and (b) paediatrics, mostly resembling puériculture. Obviously, they aimed at offering both prenatal and post-natal advice about successful methods for 'good quality' births.

The fourth and last meeting of the Working Committee during 1958 was held in exactly the same spirit as the previous one.[13] Raphael reported that he contacted the community centre 'The house of friendship' (*Εστία Φιλίας*) and agreed with its director to organise an open lecture for their audience, consisted of parents and young people. In addition, Raphael arranged lectures at the biggest textile factory in Greece, the Piraiki-Patraiki factory. Poggis, who worked at the house of friendship, made some substantial suggestions to Raphael regarding the best possible ways to disseminate eugenics. He insisted in distributing eugenics leaflets to labour groups, teachers' journals, military magazines and writing to the provincial press. Another suggestion was to contact the Archbishop Ieronymos Kotsonis, the leader of the Christian brotherhood 'Life' (*Ζωή*), because this organisation distributed leaflets of various subjects to approximately five hundred thousand Greek families. Moreover, Poggis disagreed with the talks in the provinces because he claimed that the subjects of eugenics were too complex for villagers.

The members of the Working Committee who were health professionals were asked to enlist the central eugenics arguments in everyday language in order to produce a leaflet to be distributed to workers. They also underlined the necessity of creating educational material for healthcare workers, midwives and doctors to be included in their educational programmes in universities and nurse schools.

The Working Committee drew a plan for the year 1958 and another one for 1959. The former included valuable information about the relationship between van Vleck and the Greek eugenicists. For instance, van Vleck promised to initially finance the HES with 150 dollars and later to increase his funding up to the 49% of its total budget.[14] Obviously, van Vleck encouraged the HES both morally and financially.

The most important task of the committee was to contact state authorities in order to define a specific population policy in the light of the financial, social and military situation in Greece. A specific state policy would result in a common code of practice restraining any independent private activity. As a second priority, the Working Committee suggested two possible measures aiming at 'the biological improvement of the Greek race'.[15] These were, on the one hand, the improvement of the procreation conditions and on the other hand, the avoidance of procreation in cases where diseases or negative conditions threatened the health of the parents and their descendants. They thus promoted a scheme which included both positive and negative eugenics in the form of family planning advice.

It is remarkable that their primary goal was to organise three or four lectures about the overpopulation problem by inviting experts of the field. They also planned to integrate these lectures into the context of the UN seminar on population to be held in Athens in September 1958.

The specialisation of health professionals in eugenics was also an issue that was repeated in every schedule, but also more lectures at workplaces and youth centres were included in their plans. What is more, they urged the necessity of propaganda material, such as leaflets and recorded lectures in plain language to be distributed during the conferences and lectures of the HES, at the PIKPA and Paediatric Institutions. The Working Committee claimed that these measures would be fruitful but sporadic; therefore, it was imperative to use the mass media on a regular basis. Such a task could be realised with the cooperation of the Education Service of the Ministry of Social Care. In addition, the idea of distributing a newsletter among the members of the HES was put forward in this plan of action to be distributed in February 1959. Thus, the future plans of the HES were summarised in the following: the organisation of a conference on overpopulation; the effort to attract more members; the multiplication of the publications in the press including the public talks; and the publication of a leaflet on eugenics prepared by Doctors Papaloukas and Karanastasis.

The assembly of the Executive Board accepted the plans of the Working Committee with some alterations, such as to add non-academic lectures on heredity and the organisation of talks in rural areas. Regarding the academic lectures, Louros suggested collaboration with other societies, such as the Biological Society where Pantazis was president. The Executive Board also decided instead of organising sparse scientific lectures, to try to incorporate them into academic schedules as educational courses on eugenics.

As far as the newsletter was concerned, its purpose was to revive the interest of the members of the HES and to attract new members. The newsletter was scheduled to include information about the activities of the HES, updates of similar associations abroad and international news in the field of eugenics. Unfortunately, only three issues of the newsletter have been preserved in Louros' archive: February 1959, October 1959 and June 1962.[16]

The earliest extant newsletter included a report on the lectures of 1958, which were: V. Triantafyllopoulos, 'The Pre-directed Heredity', Parnassus Hall, January 1958; K. Saroglou, 'Issues of Practical Eugenics', Parnassus Hall, February 1958; J. van Vleck, 'International Progress in the Field of Eugenics', Alexandra Maternity Hospital,

February 1958 and S. Doxiadis and M. Raphael, 'Population and Eugenics Problems from an International and Greek Perspective', *International Alliance of Women (IAW) and UNESCO International Congress*, Christian Youth Association Room, Athens, August 1958. Obviously, these lectures were attended by educated public. As agreed in their meetings, they organised three to four lectures per year, thus reviving the interest of the public in eugenics. The topics of discussion reflect their attempt to familiarise people with basic eugenics knowledge, before introducing more complex subjects.

Regarding the publications in Greek journals and newspapers, it was stated in the newsletter that the efforts of the members of the HES to popularise eugenics and birth control were very effective and a growing interest of the public in these issues was observed. For example, the journal *Ikones* (*Εικόνες*) of 18 August 1958 featured research on eugenics and birth control and an interview with Louros (Ikones 1958). It is noteworthy that many of the HES activities were published in highly respected and popular medical journals, such as *Iatriki* and *Elliniki Iatriki*, which reached a large number of Greek physicians. Equally, minutes of the conferences of the HES were widely publicised in daily newspapers, such as *Apogeymatini, Kathimerini, To Vima, Ta Nea* and others, not only distributed in the capital but also nationally. Obtaining access to the mass media of the time was undoubtedly an important advantage.

As for the international relationships of the HES, the visits of foreign experts and the donations by van Vleck and Dorothy Brush were highlighted as they were substantial contributions to the work of the HES. The fact that Van Vleck congratulated the HES on its activities and the idea of the newsletter and his promise to refer to the HES at the IPPF's Conference in Bombay in 1959 was also included in the newsletter. The section of the international news of the newsletter included: a table showing population movement in France; the falling birth rate in Japan and family planning advice in public hospitals in New York, India and Egypt. There was also a report on the seminar on population, organised by the United Nations Bureau of Social Affairs and Technical Assistance Administration in cooperation with the Government of Greece, held in September 1958 in Athens (United Nations 1959). The President of the conference was the Greek Professor Gerasimos Alivizatos. Vasilios Valaoras, a former member of the HES, represented the UN.

The second newsletter included information concerning the public lectures of the HES in 1959, as follows: in February 1959, Louros talked about problems of alcoholism under the aegis of the Hellenic Society of Anti-Alcoholism, at Parnassus Hall; in March 1959, Mrs. Olga Chysostomidou talked about problems of infancy at the house of friendship, where the audience showed particular interest in family planning issues; in May 1959, Pantazis, Vice-President of the HES, was invited by the Italian government to give a series of talks at Italian universities about 'Overpopulation as a biological problem'; in August 1959, Louros spoke in Helsinki, Finland, about Overpopulation and Birth Control where he highlighted the need for 'an international birth control, but not only regional, which would unavoidably lead to the suicide of the white race'.

Although the lectures in Greece referred to general topics, the lectures abroad, both Pantazis' and Louros' lectures, referred to overpopulation. During the same year, 1959, the HES would organise one of its most successful conferences on the subject of overpopulation. Indeed, by the end of the 1950s, it was one of the most popular subjects of discussion among eugenicists who endorsed birth control.

The success of the conference on overpopulation revived the interest of people to the activities of the HES. The fact that the HES regained its popularity in the 1960s is obvious from its newsletter of June 1962, which included a report on the General Assembly and three successful round-table public discussions. The annual General Assembly of the HES was held at Alexandra Maternity Hospital on 21 February 1962. Louros presented the work of the HES during 1961, particularly mentioning the success of the conference on Euthanasia which was held in March 1961 (Hellenic Eugenics Society 1965).

Equally important topic was the health and physical education of children on which the HES devoted two conferences in 1962. The venue of the conferences was changed from the Parnassus Hall to the more spacious 'Kentrikon' theatre. It is remarkable that both conferences were attended by the Greek Princes Peter and Michael. The subject: 'The health state of the Greek children' was discussed on 5 March 1962 and the second discussion followed two weeks later, on 19 March, with the subject: 'The physical education of the Greek children'.

The outcomes of the conference included the fact that the physical education in high schools was poor and the twenty per cent of the Greek children do not know how to swim. They also questioned the fact that the remaining eighty per cent indeed knew how to swim. They feared that in reality the percentage was much lower. They highlighted that swimming was a personal choice, whereas the physical education and exercise was a matter of the state (Kathimerini 1962). However, the conferences were concluded rather optimistically by arguing that the Greek population had greatly improved in health, robustness and mental development during the last two decades (Mesimvrini 1962; Konstantinidis 1962).

The conferences of the HES received acceptance and appreciation both from experts and the general public. Apart from the popularity of its members and guests, the success of the HES' activities was highly attributed to external support. The HES was not at all restricted to its national borders; on the contrary, its president and members enjoyed international recognition and support.

4   Collaborations in Greece

By 1954, the HES was receiving more acknowledgement from its international contacts than from its own public in Greece. However, this was soon to change. Some of the crucial events which took place during 1954 were as follows: in April the statutes were officially approved, in May–June Kanavarioti visited Britain, in August Louros was elected President of the HES and in September the World Population Conference took place in Rome. Additionally, the Alexandra Maternity Hospital was fully established in Athens in December of the same year (Adamantidou and Vantzeli). Louros became the scientific director of the hospital. In coming years, the Greek eugenics movement and family planning campaign would be associated with that institution. The IPPF's experts visited the Alexandra Maternity Hospital and praised its innovative work and modern infrastructure. During the same period, Louros was simultaneously an active obstetrician and gynaecologist, Professor of Obstetrics and Gynaecology at the Medical School at the University of Athens, Scientific Director of Alexandra Maternity Hospital and President of the HES. He thus had all the available means to disseminate eugenics in theory and practice. Furthermore, by the end of 1954 the network including the HES, the IPPF and the British Eugenics Society was well established. There were many meetings and interactions among people belonging to these institutions. Furthermore, international relationships also helped the HES to expand its work locally, too.

A critical moment was Louros' first public lecture on eugenics in front of a Greek audience, which inaugurated the HES's public activities in the country (Louros 1955). The content of the lecture was based on Vera

Houghton's recommendations,[17] such as the works of C. P. Blacker (1947), Paul Bloomfield (1948) and Cedric Carter (1952). Houghton was the Executive Secretary of the IPPF's office in London. The available information on eugenics was carefully adapted to the Greek social, political and medical model by Louros. He began the lecture by giving a definition of eugenics to the allegedly ignorant audience. He said thus:

> Eugenics (ευγονική) is the science which deals with the matter of 'good birth' (ευγονία); i.e. with the factors that improve the qualities of a race and the factors that develop these qualities to the highest level.[18]

He attributed the above definition to Galton, of course, whom he characterised as 'knowledgeable of Greece' (ελληνομαθής). He argued that eugenic practices in Ancient Greece revealed that the human need for racial improvement was not an innovative theory of the twentieth century. On the contrary, the self-preservation instinct dictated that humans pursue a better life. The choice of spouse itself stems from the human inclination to improve, because people seek the most suitable 'partner in reproduction'. What Louros was willing to say was that subconsciously people choose a partner not solely based on sentiment, but also because of his/her potential of becoming a good mother or father, both genetically and intellectually. Louros interpreted this attitude as a manifestation of eugenics which was intrinsic to human nature. The combination of hereditary predisposition, which is the genotype, and the result of the environmental influence on the genotype, which is the phenotype, was essential to eugenics. Eugenics could be achieved either by finding the optimal combination of these two parameters or by eliminating the harmful genotype. Louros explained that human should opt for the proper choice of spouses in conjunction with the amelioration of living conditions to achieve eugenics. Louros did not adopt a genetically deterministic approach, but acknowledged the environmental influence as equal factor to achieving the goal of eugenics. This view was shared by eugenicists at the time, as was mirrored in the HES conferences. Medical professionals, biologists, sociologists and economists discussed the multifactorial nature of human evolution. Living conditions, natural environment, social norms and education were some of the factors which influenced humans and affected their development, intelligence and behaviour.

Louros argued that eugenics was not an easy task to accomplish because many obstacles could render this process impossible. The greater part of the lecture was devoted to the restraining factors of eugenics application. These were categorised as moral, medical, administrative, socio-economic and political obstacles. Moral issues included inappropriate marriages, given that few people were suitable for marriage and reproduction. Louros acknowledged a gradual phenotypic decadence in the society of his time by a wide moral degeneration caused by alcoholism, prostitution, drug addiction, lack of respect and criminality which shook the foundations of society and democracy.

Medical problems were equally important and very difficult to deal with. The core problem was the difficulty with categorising people based on their suitability for reproduction due to each individual's unique combination of traits. Therefore, any recommendation for suitability for procreation was neither achievable nor effective. Furthermore, there were as many scientific difficulties for birth control and the limitation of large families as there were for the diagnosis and cure of sterility. The medical resources were relatively poor at that time, and people were reluctant to trust them. Louros agreed with Soranus of Ephesus' (ca. 98–138 A.D.) proverb that: 'Non-conception is preferable to abortion'.[19] However, he argued that 'non-conception' should not be understood as forced sterilisation. He was extremely critical of both forced sterilisation and abortion.

In the medical context, Louros regarded preventive medicine as absolutely necessary for every citizen. He believed that the profit from the limitation of diseases would outweigh the additional investment in the implementation of preventive medicine.

Louros also argued that one of the most important socio-economic problems was the disequilibrium between the small and wealthy families, in contrast to the large and poor ones. Although wealthy children were not necessarily more competent, they had the available means to become so. However, he argued that often the leaders of their society came from poor backgrounds. In this context, Malthus' population theory was mentioned and supported by Louros, insofar as to social protection from the negative consequences of overpopulation. The issue of Greek demography could not be overridden by Louros. According to the biostatistician Valaoras, the death rate had fallen in Greece after the Second World War, resulting in population growth. Louros briefly claimed that if the Greek population continued to increase, the Greek economy would be unable to sustain it. At the same time, however, birth control was forbidden in Greece by religious and political bodies.

Despite these obstacles, Louros urged the immediate need for eugenic policies. This lecture gave him the opportunity to present his eugenic viewpoint and to try to convince the audience that eugenics was essential for the Greek society. Some possible ways to overcome the difficulties of the application of eugenic policies were the study of heredity, the implementation of methods for mental and psychological calculation of the prospective parents, the study of deviation from normality, the study of the environmental influence, biostatistics, geophysics, financial eugenic views and evaluation of the demographic problem. The crucial issue was the influence of the genotype. According to Louros, the optimal solution was preventing parents with defective genes from reproduction. In addition, the improvement of nutrition, housing and education would improve the phenotype, the manifestation of the genotype. Last but not least, the family planning techniques should be implemented in order to avoid large families and overpopulation.

In conclusion, Louros admitted that every social change could only be realised by political initiatives. The newly founded HES would undertake the responsibility for informing and educating the political leaders about the science of eugenics. Louros called the audience to help the HES' efforts by participating in its struggle for eugenics research, education and ultimately, human survival.

People from the IPPF showed particular interest in the success of this lecture, after Kanavarioti's report on 14 March 1955. Among the first who responded was Rotha Peers, who was thrilled by the fact that the lecture attracted as large an audience as 800 people.[20] Houghton mostly praised Kanavarioti's work on preparing Louros' lecture. During the preparation of the lecture, Houghton expressed her support to Kanavarioti by writing that the establishment of the HES was primarily due to her efforts. According to Houghton, without Kanavarioti's inspiration and persistence, the HES would have never been established. Apart from Houghton and Peers, Clarence J. Gamble commented on the successful lecture, too. Houghton urged Kanavarioti to inform Dorothy Brush, the editor of the journal *Around the World News on Population and Birth Control*, of the success of Louros' lecture. Houghton assured Kanavarioti that Brush would be delighted by the news. Indeed, Brush included a section for Greece in the journal as follows:

> The Hellenic Eugenics Society, located in Athens, recently made its first appearance in public with three important lectures. This contribution with

the pioneer organisation met with an unexpectedly wide response: every seat was filled in Parnassus Hall, the largest auditorium in Athens. [...] We congratulate Mrs. Maro Kanavarioti and her associates who have worked patiently and persistently to bring the knowledge and recognition of planned parenthood to Greece.[21]

The lecture was particularly fruitful in terms of gaining non-physician supporters of the eugenics movement. The HES was indeed a unique association in this respect. For example, Georgios Adamopoulos, an Astronomer and Director of the Astronomical Institute of Athens sent a congratulatory letter to Louros right after the lecture.[22] Adamopoulos claimed that a speech, such as the one Louros delivered, was deemed necessary to inform the Greeks about the dangers of the uncontrolled population growth. Adamopoulos continued with a brief analysis of the population problem and Malthus' theory. He considered the eugenic view of the creation of genetically perfect man as completely utopian; simultaneously suggesting the constraint of uncontrollable population growth as the only solution. Finally, he requested from Louros to be included in the HES as a regular member. Louros positively responded three days later.[23] Adamopoulos soon became a member of the Executive Board of the HES and then fellow of the British Eugenics Society.

In addition, Michael Goutos, Vice-President of the Greek Social Insurance Institution (IKA), was delighted by Louros' lecture and suggested the publication of the text in the, then new, journal of the IKA. He also suggested that the lecture would be translated in English by the Department of Foreign Publishing at Yale University and distributed in the USA too. Another suggestion was the inclusion of the HES, represented by Maro Kanavarioti, in a newly formed Union for the Study of Social Protection Issues (Σωματείο Μελέτης των Θεμάτων Κοινωνικής Προστασίας). Moreover, he asked Louros' permission to publish his paper given in the Obstetrics and Gynaecology Conference in Geneva in July 1954. In response, Louros agreed to the publication of his papers and Kanavarioti's participation in the Union for the Study of Social Protection Issues.

After the success of the lecture, Louros opened up the HES to more people by inviting important people outside the medical field to join the HES as a way to popularise its work and expand its network. Among them were: S. Kalliafas, working at the Laboratory of Experimental Pedagogy[24]; I. Karmiris, Royal representative at the Holy Synod of the Greek Orthodox Church[25]; and P. Anagnostopoulos, Professor of Horticulture at the Higher School of Agriculture.[26]

Starting from Louros' lecture, the HES gradually developed connections with institutions, organisations, unions and associations to promote the dissemination of eugenics in Greece. Indeed, the HES shared members and ideals with similar Greek associations. Many members of the HES held important political, social and professional posts that made those connections much easier to be accomplished.

For example, the HES was in contact with the National Union of Sanitary Education (NUSE). The NUSE was the representative of the Union Internationale d' Education Sanitaire, a non-governmental organisation founded in France in 1951 (Population 1951). The NUSE was founded in 1954 (National Union of Hygienic Education 1960). Georgios Pangalos, Professor of Hygiene and President of the NUSE, sent a personal letter to Lina Tsaldaris, President of the PIKPA, informing her of the approval that the NUSE gained from the Court of First Instance and the text of its statutes.[27] The main purpose of the letter was to ask her about any suggestions for possible, prospective members for the NUSE. He specified that it was not necessary for them to be physicians. Pangalos also mentioned that the announcement of the foundation of the NUSE was included in the *Bulletin de Liaison et d' Information* (1954) published by the Union Internationale pour l' Education Sanitaire de la Population, whose Greek representative then became the NUSE. While the HES was primarily associated with the USA and the UK, the NUSE was connected with a French, later international, institution.

The text of the statutes included the following articles which defined the aims and composition of the NUSE.[28] The first article defined that the NUSE will be based at the School of Hygiene in Athens. The second article outlined its aims which were the public dissemination of hygienic knowledge and preventive medicine in cooperation with public services and private organisations, the coordination of every private undertaking towards this target and the effort to make the NUSE equivalent to other countries' representatives of the Union Internationale d' Education Sanitaire. Similar to the aims of the HES, educating the public was of utmost importance. Both institutions were established by well-known professionals and academics in order to alert the wider public about the importance of preventive health, hygiene and eugenics. Often, it became difficult to identify when the threshold between preventive health and eugenics in matters of reproduction and disease prevention was crossed by these experts. Furthermore, the desire to cooperate with state authorities was a prerequisite for the function of the NUSE too.

However, the third article of the statutes referred to the fact that the NUSE did not belong to any of the public services but included members who worked in the public sector, members who worked in the private sector and individuals who were interested in its aims. The statutes were officially deposited in the Court of First Instance on 5 June 1954.

As already mentioned, in 1952, the innovative service of the Ministry of Social Welfare for public education in matters of hygiene and preventive medicine gave physicians and healthcare professionals the opportunity to develop activities under the auspices of the Ministry. In just a few years, a network of academics and health professionals who aimed at the amelioration of Greek society was created with the support of the state. It was in this context that both the NUSE and HES were established in the early 1950s. Both lists of members included not only eminent physicians and medical academics, but also people who worked at public institutions such as the PIKPA, the School of Hygiene, the university maternity hospitals, Alexandra and Marika Iliadi, various Schools of Nurses and the departments of the Ministry of Social Welfare. Notably, the director of the newly founded Education Service of the Ministry of Social Welfare, Kapalas, was also founding member of the NUSE.

The list of the founding members reveals important information about the relationship between the NUSE and the HES, but also about the overall situation in Greece regarding hygiene and eugenics. First of all, the NUSE was established in approximately the same time as the HES. Secondly, ten out of thirty-five founding members were also members of the HES. These were: Lina Tsaldaris; Konstantinos Moutoussis, Konstantinos Charitakis, Nikolaos Louros, Konstantinos Choremis, Georgios Krimpas, Antonios Papadakis, Michael Goutos, Vasilios Malamos and Gerasimos Alivizatos. Remarkably, leading members of the HES, such as Louros, Moutoussis, Choremis, Papadakis and Goutos, participated in both institutions. The connection between the two institutions culminated in the organisation of joint public discussions from 1955 to 1956.

Furthermore, the membership of ten women in the NUSE should not be overlooked, all of them holding leading positions, with Tsaldaris having the uppermost at the directorship of the PIKPA. Both the NUSE and the HES had female members and often invited female scholars to participate in their meetings. Added to this, the HES was represented abroad by Maro Kanavarioti for many years. Medical circles, albeit male dominant, included many women. These women, not only were not underestimated by their colleagues, but were recognised as valuable contributors to the progress of medicine in Greece. Furthermore, Louros always mentioned the importance of the female nurses in gynaecologist's work and their unique ability to reach female patients (Louros and Kairis 1951).

As was mentioned in the NUSE's publications, its purpose was the 'public propaganda of crucial elements of hygiene and preventive medicine'. In this context, they persuaded the National Radio Institution (*Εθνικό Ίδρυμα Ραδιοφωνίας*) to record more than fifty short lectures on various issues of hygiene. In a period when television was essentially non-existent in Greece, the majority of the public were radio listeners and broadcasting was the most popular medium of information. Indeed, these lectures were very informative and simple, in order to be understood by every listener. Among the speakers were members of the HES, such as Georgios Pournaropoulos and Theodoros Zavitsanos, who talked about school hygiene and accidents, respectively.

However, most of the lectures were delivered by the NUSE's president, Georgios Pangalos. He paid particular attention to the prevention of diseases, such as tuberculosis, and did not hesitate to say that the transmission of diseases not only was a moral sin, but also a crime. He called patients 'useless and dangerous individuals, who were at the same time a financial burden to society'. He also claimed that those who suffered from diabetes should avoid procreation.

His most radical views on eugenics were revealed in his last recorded speech, under the title 'Heredity', which was eventually not approved by the National Radio Institution, and therefore, never broadcasted on air but was published. First of all, Pangalos considered the introduction of premarital health certificate to be useless, because most of the people were not scrupulous enough to decide not to procreate should their partner be unable to have healthy children. In this context, he attacked mothers with tuberculosis and alcoholic fathers, whose attitude was equated with infanticide. Moreover, he referred to the science of eugenics, which was the most appropriate way to study how to avoid defective descendants. Pangalos argued that people should be educated by eugenic studies because public health could only be protected by proper education. Moreover, Pangalos was in favour of the compulsory sterilisation of criminals, drug addicts, perverts, epileptics and psychopaths.

He supported state intervention in the sterilisation of these people using painless medical procedures without considering legal implications. According to Pangalos, individual freedom should be sacrificed for the sake of society.

Pangalos strongly supported extreme eugenic measures, contrary to most members of the HES, who were against forced sterilisation. However, there were a series of lectures, organised by both the NUSE and the HES, which took place at the premises of the Christian Youth Union (*Χριστιανική Ένωση Νέων*) in Athens. Some of the lectures were 'Protection of motherhood', delivered by Nikolaos Louros (18 November 1955); 'Heredity and eugenics of psychological illnesses', delivered by Konstantinos Konstantinidis (16 December 1955); 'Practical application of heredity', delivered by Panayiotis Panayiotou (27 January 1956); 'The psychological needs of newborns', delivered by Spyros Doxiadis (24 February 1956); 'Eugenics in flora', delivered by Dimitrios Panos (30 March 1956) and 'General principles of eugenics', delivered by Konstantinos Saroglou (20 April 1956).

By the time UNESCO organised a conference on Dissemination of Science convened in Madrid, 19–22 October 1955. Georgios Pantazis, a Greek Professor of Biology at the University of Athens and Vice-President of the HES, represented Greece. There, he referred to the HES' role in the popularisation of science in the country. Among his recommendations for ways of disseminating science, such as broadcasting, newspapers and periodicals, he wrote that private societies, such as the HES and the NUSE organised lectures of a more technical character for the general public (Pantazis 1955). Although there is no evidence for the continuation of the collaboration between the NUSE and the HES, in the following years, the HES continued to organise conferences and symposia annually until the 1980s.

The HES was also in contact with the PIKPA primarily due to Konstantinos Saroglou, who was the Medical Director of the PIKPA and confidant of Lina Tsaldaris, its president, was one of the very active, founding members of the HES and the NUSE. In particular, he was a member of the Executive Board of the HES from 1954 to 1967. The HES admired the work of the PIKPA and cooperated with it. Saroglou, of course, was the link between the two.

Unlike the NUSE and the HES, the PIKPA's activities were not theoretical but practical. It was established at the beginning of the century, based on the fundamental principles of solidarity and volunteerism.

Its work for the protection of mothers and children was the most fruitful in the country. Vaccination, communal meals, free medical examination and shelter for single mothers and their children were some of its activities. The PIKPA was one of the leading institutions in Greece that played an important role in the protection of mothers and children. During the last period of the Liberal government (1928–1932), the Greek state began to contribute funding to its activities. The PIKPA had branches in many different regions, both in urban and rural Greece.

Lina Tsaldaris participated in the first meeting of the HES at the AMA but did not attend the following meetings because of her large workload, particularly during the period when she was Minister of Social Care, from 29 February 1956 until 5 March 1958. However, Tsaldaris was on the list of the IPPF Honorary Associates, representing Greece together with Louros. Tsaldaris was a politically and socially influential person, with great experience in matters of maternal and infant care. She became the first female minister in the Greek Parliament. She participated in numerous conferences, both in Greece and abroad, regarding the protection of women and children; later she became a member of the Greek Delegation to the UN and officer liaison with UNICEF for Greece. By the 1960s, PIKPA was a well-organised and functioning institution under the leadership of Tsaldaris. The PIKPA was the instrument through which she organised her social work. In one of her letters to the UN, she described the PIKPA as 'the only official body for infantile and maternal protection in Greece' and summarised its activities in two categories.[29] Firstly, they provided families with assistance and protection by organising children camps; centres for milk distribution for preschool children and pregnant women; distribution of baby linen, clothes and shoes; material aid; adoptions and sponsorships. Secondly, they focused on preventive medicine and maternal and child hygiene services, such as prenatal consultation, consultation for proper nutrition and training of qualified personnel. They also supported the establishment and organisation of hygienic centres; dispensaries; polyclinics; mobile dental clinics; preventoria; sanatoriums and rehabilitation centres for disabled children.

In her report, Tsaldaris assured the UN that the PIKPA was a respectful organisation which needed more buildings to host its services and renovation of some buildings destroyed by the wars and the German occupation. Indeed, the PIKPA was a unique institution for child and maternal protection and care. Apart from Tsaldaris who headed the institution and disseminated its activities both abroad and locally, most of the people who worked there were both volunteers and high-qualified, such as Dr. Tsakos, a physician trained in the USA.

5   Eugenics and Politics

The HES managed to be linked with politics on many occasions and in different ways. The meaning of the phrase 'relation with politics' is defined as any linkage with the government in power, the Royal Family of Greece and state authorities in general.

On the theoretical level, from the very beginning of its foundation, the HES aimed at cooperation with the state. This was justified by the reference to the relationship with the state among the aims in the official statute and the many times that it was mentioned in their meetings. The ultimate target of the HES's activities was to transform the outcomes of its eugenics studies to legislation. As was referred during meetings and conferences, the work of the HES was geared, on the one hand, towards the dissemination of eugenics to the public, and on the other hand, to the lobbying of each government to implement eugenic policies. As Valaoras underlined in the first meeting, governmental action for the elevation of health level and motherhood protection was inadequate because government officials were unaware of eugenics. Thus, the purpose of the HES was to fill this gap and inform the state about eugenics. Furthermore, the recent past of the Second World War and the Civil War was taken into consideration by post-war eugenicists who wanted to lead the social reconstruction of the country. Notwithstanding their desire to play a crucial role to this end, they would not replace politicians. They would present the Eugenics Society as a purely scientific association, without linkage to a particular political party in order to be an independent organisation. Furthermore, the eugenics movement aimed at popularising eugenics to all people regardless of social class or political belief, because the ultimate goal was the regeneration of the society as a whole. Hence, there were members of the HES, who were ministers or secretaries in the Ministry of Health and/or Education or they were familiar with members of the government or the Royal Family.

First of all, Athanasios Mantellos, who was the first President of the HES, became General Director at the Ministry of Social Care. Here, it has to be repeated that the Ministry of Health changed to a variety of names, such as Ministry of Social Care; Health; Health and Hygiene;

Hygiene, Social Care and Perception; State Hygiene and Perception, remaining the same service throughout, however.

Nikolaos Louros was one of the most politically involved presidents, even if he declared himself as politically neutral. In fact, his friendship with politicians and the Royal Family is attributed to his father. Louros' father, Konstantinos Louros, was a prominent gynaecologist and the gynaecologist of the Royal Family; therefore his son had connections with them from an early age. In his autobiographical book *Yesterday* (Louros 1980), Louros referred to the summers he spent in Tatoi, the Royal residence, as well as his familial excursions in Kifisia, a suburb of Athens, where most of the politicians and scholars lived. Louros' father was also deputy of the People's Party under Panayis Tsaldaris. Moreover, he was Secretary at the Ministry of Health during the short period of a month from 10 October 1935 until 30 November 1935. Following his father's steps, Louros became a respected obstetrician and gynaecologist and succeeded him in the service to the Royal Family. In 1939, Nikolaos Louros and Kurt Warnerkros assisted the birth of the future Queen Sophia of Spain and in 1940 the birth of her brother, future King Konstantinos of Greece, receiving medals from the Royal Family on both occasions. In turn, a representative of the Royal family, such as Prince Peter and Prince Michael, often attended conferences of the HES during Louros' presidency. Regarding governmental positions, Louros participated in the Government of National Unity (established in 1974) under Prime Minister Konstantinos Karamanlis, when he became Minister of National Education and Religion. Moreover, he participated in two very important state committees; the Committee for Education, where Achilleas Gerokostopoulos was the president and Louros was one of the six members and the Committee for Matters of Social Insurance, where Louros was the president. At that time, he had already published his work on the sanitarian organisation of the country (Louros 1945). Moreover, Louros became Minister of Education for a short period between July and November 1974 and a member of the National Hygiene Council.

During the government of Konstantinos Karamanlis, Spyros Doxiadis, one of the founding members of the HES and its president in 1973, was involved in politics twice. In the first instance, he became Minister of Social Services for only two months (October–November 1974). A few years later he became Minister of Health (November 1977–October 1981). Given that the Ministry of Health took several names during the twentieth century, but remaining the same service, Doxiadis was

Minister of Heath for the longest time period, in total 48 months and 5 days (Dardavesis 2008). Apostolos Doxiadis, father of Spyros Doxiadis, a pre-war eugenicist and himself also Minister of Health from 17 September 1922 until 12 March 1924 and Secretary at the same Ministry from 25 August 1928 until 7 June 1929. Generally, the Doxiadis family was renowned in Greece due to the professional success of its members such as the aforementioned ones and the internationally famous architect and urban planner Konstantinos Doxiadis who participated in one conference of the HES: 'Environment and Survival' on 8 April 1971.

Gerasimos Alivizatos, a member of the HES, held the post of Secretary in the Ministry of Health from 5 August 1936 until 12 December 1938. Moreover, Lina Tsaldaris was Minister of Social Care during the period from 29 February 1956 to 5 March 1958. Another member, Evangelos Papanoutsos, a theologian and pedagogue, was appointed General Director in the Ministry of Education from 1944 to 1946. Later, in 1950, he became General Secretary in the same Ministry and he also held the same position in 1963–1964.

The official political posts held by members of the HES and most notably by its presidents are only examples of their wider involvement in the politics of the country. It is important to mention that the relationship with politics was mutual. Many politicians, including members of the Royal Family, attended or delivered speeches in the conferences of the HES thus overtly endorsing eugenics. Due to the fact that the majority of the members were scholars, academics and renowned physicians, their contact with the sociopolitical elite of the country was guaranteed.

NOTES

1. Athens Medical Association Archive 1950–1952.
2. Hellenic Eugenics Society. 1953. Proceedings of the Meeting 29 March 1953. *Nikolaos Louros Papers and Archive.*
3. Danopoulos to Kanavarioti. 1953. *Nikolaos Louros Papers and Archive.*
4. Kanavarioti to Whelpton. 1953. *Nikolaos Louros Papers and Archive.*
5. Ibidem.
6. Kanavarioti to Vogt. 1953. *Nikolaos Louros Papers and Archive.*
7. See Chapter 5.
8. Newspaper extract. Ta Nea, 25 January 1955. *Nikolaos Louros Papers and Archive.*
9. Hellenic Eugenics Society. 1958. Minutes of the executive board meeting. *Nikolaos Louros Papers and Archive.*

10. The Margaret Sanger Papers Project. New York University.
11. Hellenic Eugenics Society. First meeting of the Working Committee 8 January 1958. *Nikolaos Louros Papers and Archive.*
12. Hellenic Eugenics Society. 1958. Third Meeting of the Working Committee of the Hellenic Eugenics Society. February 5. *Nikolaos Louros Papers and Archive.*
13. Hellenic Eugenics Society. 1958. Fourth Meeting of the Working Committee of the Hellenic Eugenics Society. February 12. *Nikolaos Louros Papers and Archive.*
14. Hellenic Eugenics Society. 1958. The annual plan of the Hellenic Eugenics Society. *Nikolaos Louros Papers and Archive.*
15. Hellenic Eugenics Society. 1958. The plan of action of the Hellenic Eugenics Society 1958–1959. *Nikolaos Louros Papers and Archive.*
16. Hellenic Eugenics Society. 1959. Newsletter of the Hellenic Eugenics Society. *Nikolaos Louros Papers and Archive.*
17. Houghton to Kanavarioti. 1954. *Nikolaos Louros Papers and Archive.*
18. See a more detailed definition of the Greek words ευγονία and ευγονική in the first chapter of this book.
19. In Greek: Το μη συλλάβειν πολύ μάλλον συμφέρει του φθείρειν.
20. Peers to Kanavarioti. 1955. *Nikolaos Louros Papers and Archive.*
21. Around the World News on Population and Birth Control 35. 1955. *Dorothy Hamilton Brush Papers.*
22. Adamopoulos to Louros. 1955. *Nikolaos Louros Papers and Archive.*
23. Louros to Adamopoulos. 1955. *Nikolaos Louros Papers and Archive.*
24. Kalliafas to Louros. 1955. *Nikolaos Louros Papers and Archive.*
25. Karmiris to Louros. 1955. *Nikolaos Louros Papers and Archive.*
26. Anagnostopoulos to Louros. 1955. *Nikolaos Louros Papers and Archive.*
27. Pangalos to Tsaldaris. 1954. *Lina Tsaldaris Archive.*
28. The Statutes of the National Union of Sanitary Education. 1954. *Lina Tsaldaris Archive.*
29. Tsaldaris, Lina. 1951. Informations Relatives au questionnaire resultant de la resolution 390 D (XIII) du Conseil Economique et Social des Nations Unies adaptée le 9 Aout 1951. *Lina Tsaldaris Archive.*

REFERENCES

About Greek Children: The Public Discussion of the Eugenics Society. 1962. *Kathimerini*, March 6.
Adamantidou, Triantafyllia, and Kiriaki Vantzeli. The History of the Alexandra Maternity Hospital. http://www.hosp-alexandra.gr/index.php?option=com_content&view=article&id=84&Itemid=67. Accessed 11 Jan 2012.
A Pill Against Malthus' Prophecy. 1958. *Ikones* 147: 30–33.

Athens Medical Association Archive. Minutes: vol. December 1950–March 1952 and vol. March 1952–March 1955.

Blacker, Carlos P. 1947. What Is Eugenics? *The Eugenics Review* 39 (2): 56–58.

Bloomfield, Paul. 1948. The Eugenics of the Utopians. *Paper Read to the Eugenics Society.*

Burnova, Evgenia. 2005. Deaths from Hunger: Athens in the Winter of 1941–1942. *Archeiotaxio* 7: 52–73.

Carter, Cedric. 1952. Eugenics in the Prevention of Hereditary Disease. Reprint from *The Medical Press.*

Clarence Gamble Papers, Series: III. Countries Correspondence and Records, 1927–1965, box 77 (1207–1215) HMS c23. Harvard Medical Library, Francis A. Countway Library of Medicine, Boston, MA.

Création de l'Union Internationale pour l'Éducation Sanitaire Populaire. 1951. *Population* 6 (4): 733. http://www.persee.fr/web/revues/home/prescript/article/pop_0032-4663_1951_num_6_4_2676. Accessed 11 Apr 2015.

Dardavesis, Theodoros. 2008. The Historical Course of the Ministry of Health in Greece, 1833–1981. *Iatriko Vima* 115: 50–61.

Darwin, Leonard. 1921. The Aims and Methods of Eugenical Societies. *Science* 54 (1397): 313–323.

Deltion Iatrikou Syllogou Athinon. 1952a. Editorial 10 (5): 4.

Deltion Iatrikou Syllogou Athinon. 1952b. Announcement: Series of Lectures by the Crusade of the Scientific and Social Organisations for the Psychological, Mental and Physical Health of Greek People 10 (5): 20–21.

Dorothy Hamilton Brush Papers. Northampton, MA: Sophia Smith Collection, Smith College.

Doxiadis, Spyros. 1953. The Impact of British Nationalised Medicine to the Physician and the Patient. *Deltion Iatrikou Syllogou Athinon* 11 (1): 12–14.

Durand, John D. 1964. Pascal Kidder Whelpton (1893–1964). *Population Index* 30 (3): 323–328.

Editor. 1954. Notes and Memoranda. *Eugenics Review,* 122.

Galton, Francis. 1909. *Essays in Eugenics.* London: The Eugenics Education Society.

Gardikas, Konstantinos. 1952. Medical Education in England. *Deltion Iatrikou Syllogou Athinon* 10 (10–12): 24–26.

Hellenic Eugenics Society. 1965. *Public Discussions.* Athens: Parisianos.

Hellenic Eugenics Society. 1977. *Public Discussions.* Athens: Parisianos.

Hionidou, Violeta. 2006. *Famine and Death in Occupied Greece, 1941–1944.* Cambridge: Cambridge University Press.

Konstantinidis, Th. I. 1962. Our Race Became Robust and Beautiful. *To Vima,* March 6.

Lina Tsaldaris Archive. Athens: Konstantinos Karamanlis Foundation.

Louros, Nikolaos. 1945. *The Health System of the Country: A Plan.* Athens: K. Papadogiannis.

Louros, Nikolaos. 1955. Eugenics: An Appeal. *Elliniki Iatriki* 24 (1): 289–296.

Louros, Nikolaos. 1980. *Yesterday: Memories, Impressions and Reflections in Seven Acts.* Athens: n.p.

Louros, N., and N.M. Kairis. 1951. Some Aspects of Midwifery in Greece. *The British Medical Journal* 2 (4723): 110–111.

McDonough, Megan. 2013. Lelia K. Washburn, History Professor. Obituaries. *The Washington Post.* www.washingtonpost.com. Accessed 23 Feb 2014.

Moisidis, Moisis. 1925a. *Eugenics and Marriage; Eugenics and Puericulture in Ancient Greece: Contribution to the History of Puériculture.* Athens: n.p.

Moisidis, Moisis. 1925b. *Woman: Hygiene of Marriage and Married Woman.* Alexandria: n.p.

Moisidis, Moisis. 1928. *Abortion in Ancient Greece: Forensic, Clinical and Pharmacological Study.* Athens: n.p.

National Union of Hygienic Education. 1960. *58 Lectures on Hygiene.* Athens: Yiotis.

Nikolaos Louros Papers and Archive. N. Louros Foundation, Division of History of Medicine, Faculty of Medicine, University of Crete.

Pantazis, George. 1955. Notes of Present Facilities for the Dissemination of Science in Greece. UNESCO Conference on Dissemination of Science, 19–22 October 1955, Madrid. UNESCO/NF/DIF/13.

Quine, Maria-Sophia. 1996. *Population Politics in Twentieth-Century Europe.* London and New York: Routledge.

Schneider, William H. 1990. *Quality and Quantity: The Quest for Biological Regeneration in Twentieth-Century France.* Cambridge: Cambridge University Press.

Spelioti-Bazina, Popi. 1952. *The Problems of a Working Woman.* Athens: n.p.

Stampolitis, Nikolaos. 1955. Union Recognition: Hellenic Eugenics Society. *Apogeymatini*, March 1.

Stavropoulos, Alexandre M. 1970. *Bilan analytique et clinique du Centre Experimental de Consultations Prémaritales et Conjugales de la Société Hellénique d' Eugénisme a Athènes.* Louvain: Université Catholique de Louvain.

The Greek Race Is Improving. 1962. *Mesimvrini*, March 6.

The Margaret Sanger Papers Project. New York University. http://www.nyu.edu/projects/sanger/aboutms/organisation_ippf.php. Accessed 9 Oct 2014.

Turda, Marius. 2007. The First Debates on Eugenics in Hungary, 1910–1918. In *Blood and Homeland: Eugenics and Racial Nationalism in Central and Southeast Europe 1900–1940*, ed. Marius Turda and Paul J. Weindling, 185–222.

Turda, Marius. 2014. *Eugenics and Nation in Early 20th Century Hungary.* Basingstoke: Palgrave Macmillan.

Union Internationale pour l' Éducation Sanitaire de la Population. 1954. *Bulletin de Liaison et d' Information*, May/June.

United Nations. 1959. *Seminar on Population Studies in Southern European Countries, Athens, 15–16 September 1958.* New York: United Nations.

Valaoras, Vasilios G. 1946. Some Effects of Famine on the Population of Greece. *The Millbank Memorial Fund Quarterly* 24 (13): 215–234.

Valaoras, Vasilios G. 1953a. Our Hygienic Problem. Men and Production: The Fundamental Problem of Greece. *Deltion Iatrikou Syllogou Athinon* 11 (2): 5–6.

Valaoras, Vasilios G. 1953b. Health and Safety of the Youth. *Deltion Iatrikou Syllogou Athinon* 11 (4): 33–36.

World Health Organisation. 1949. Second World Health Organisation Assembly. Proposed Change in Date for World Health Day. http://apps.who.int/iris/handle/10665/98718. Accessed 7 Nov 2013.

Wright Gillham, Nicholas. 2001. *A Life of Sir Francis Galton: From African Exploration to the Birth of Eugenics.* Oxford: Oxford University Press.

CHAPTER 4

The International Network

While the Hellenic Eugenics Society's public engagement in Greece was rather slow, foreign contacts were actively developed from as early as 1952, when Whelpton visited Athens. The HES was established much later than its equivalents elsewhere in Western Europe and the USA. However, most of its members, and in particular its president, Nikolaos Louros, lived and studied abroad for many years. Valaoras, for instance, lived in the USA, while Spyros Doxiadis practised medicine in Britain from 1945 until 1952. Furthermore, between 1952 and 1953, three different articles on English medical practice were published in the *Bulletin of the Athens Medical Association*. These were Konstantinos Gardikas' overview of medical education in England (Gardikas 1952); Spyros Doxiadis' discussion of the effects of British nationalised medicine on doctors and patients (Doxiadis 1953); and Nikolaos Rasidakis' examination of the English psychiatric system (Rasidakis 1953). Connections with Britain and the USA were closer than with other Western countries and they are fully documented by the frequent correspondence between the HES and institutions like the IPPF and the British Eugenics Society (hereafter BES). Taking into consideration that the Eugenics Societies of Britain and the USA were active until 1968 and 1973 respectively, it was reasonable to develop relations with these two.

© The Author(s) 2019
A. Barmpouti, *Post-War Eugenics, Reproductive Choices and Population Policies in Greece, 1950s–1980s,*
https://doi.org/10.1007/978-3-030-03568-6_4

1   THE INTERNATIONAL PLANNED PARENTHOOD FEDERATION

A regular correspondence with foreign eugenicists was maintained mostly between 1953 and 1955, whereas interaction with people and institutions in Greece was more frequent after 1955. In both cases, it was Maro Kanavarioti who developed relationships not only through correspondence but also through her personal visits to Britain and further afield. As has already been mentioned, the official approval of the statutes of the HES in April 1954 can be described as a pivotal moment in the history of eugenics in Greece. Before that, Kanavarioti and other Greek eugenicists were more interested in receiving guidance from foreign institutions. As her letter to Whelpton reveals, Kanavarioti had established contacts with eugenicists and demographers overseas by 1952.[1] Tellingly, in the mid- and late 1950s, Kanavarioti, Valaoras (Editor 1959) and Adamopoulos (The Eugenics Society 1957) also became fellows of the British Eugenics Society. The establishment of a Eugenics Society in Greece was, therefore, inextricably linked with the relationships that had already been developed with eugenicists in Britain and elsewhere. Kanavarioti and the HES were also in close contact with Margaret Sanger's Research Bureau and the IPPF—another branch of Sanger's activities in family planning.[2] Foreign contacts included key persons of these organisations, such as Pascal K. Whelpton, Clarence J. Gamble, Abraham Stone, William Vogt, Joseph Van Vleck, Dorothy Brush and Vera Houghton. These organisations wanted to include Greece among their partner countries. Since there was no official association dealing with eugenics and birth control before the creation of the HES, its creation became an opportunity to expand these foreign organisations' activities in this country as well. As a result, Kanavarioti became member of the Governing Body of the IPPF in 1954,[3] and the HES was made the representative of the IPPF in Greece.

The IPPF was founded in the context of the Family Planning Association's (FPA) third International Conference on Planned Parenthood, convened in Bombay in 1952. The FPA, formerly the National Birth Control Association, was an alliance of many groups that were interested in birth control and attached to the Walworth Centre, which in turn was founded in London by the Malthusian League (Houghton 1961). Preceding the IPPF, the International Committee on Planned Parenthood (ICPP) was a committee with two representatives from Britain, two from the Netherlands, two from Sweden and three from the USA. The ICPP was primarily funded by the Brush Foundation for Race Betterment. The BES provided the IPPF with free accommodation for its activities at its premises at 69 Eccleston Square, London (Houghton 1962). Although its funding came from an American institution, it was Sanger's decision to headquarter the organisation in London (Blacker 1964). The official foundation of such an international organisation as the IPPF was the result of the neo-Malthusian movement, empowered by the efforts of Margaret Sanger and Marie Stopes[4] to globally disseminate birth control practices. Instead of 'neo-Malthusianism' and 'birth control', the terms 'family planning' and 'planned parenthood' were adopted; this eugenic language was well chosen and seemed ethically more neutral.

The birth control movement greatly benefited from the Brush Foundation. Dorothy Brush's father-in-law, after his son's death, established the Brush Foundation aiming at funding research on birth control. In 1952, the Brush Foundation undertook the publication of the journal *Around the World News on Population and Birth Control* (later *International Planned Parenthood News*). Dorothy Brush was its editor and the advisory council included Margaret Sanger, William Vogt and Abraham Stone; all actively engaged with the IPPF and Margaret Sanger Research Bureau.

In 1954, Margaret Sanger was still President of the IPPF; Shrimati Dhanvanthi Rama Rau from India its Chairman; and Carlos P. Blacker was Vice-President while simultaneously carrying out his duties as Secretary of the BES. The IPPF's regional department concerned with the Europe, the Near East and Africa was established in 1952. Nancy Raphael was the Regional Honorary Secretary. On 18 February 1954, Raphael contacted Kanavarioti to ask for a list of names of eminent Greeks who sympathised with the work of the IPPF in order to include them in its list of Honorary Associates. In her letter to Kanavarioti, Raphael explained that it was important for the IPPF to include in the list of its Honorary Associates names of prestigious people from each partner country. The enlisted people had no obligation to the organisation, provided that they embraced the IPPF's ethos; they would only allow the IPPF to use their name.[5] The enlisting of Honorary Associates was a method by which the IPPF attempted to appear more credible and prestigious. The Greek names listed were those of Nikolaos Louros and Lina Tsaldaris.

A few years later, in September 1955, a letter addressed to the HES was sent by the IPPF in London having the same purpose. Although not signed, the sender was allegedly Vera Houghton, who undertook the preparation of the Tokyo conference to be held in October 1955.[6] The main purpose of the letter was to appeal for sponsorship for the Tokyo conference. As revealed by its content, the above-mentioned Louros and Tsaldaris had given their names since the Bombay conference in 1952. The sender of the letter highlighted the problem of the high rate of abortions in Japan which is why the IPPF chose Tokyo to convey the fifth International Conference. Opting to change the situation from abortion to contraception, they used every possible way to popularise their work. Among their activities was to make a list of sponsorship for the conference. Thus, they hoped to gain local acceptance. Obviously, the same letter was distributed to anyone who put their name to the list of sponsors of the IPPF's conferences. It was a typical procedure. The issue of sponsorship was brought up again by Houghton in two letters, one dated 9 September 1955, and another one on 12 July 1956, when she prepared the report of the Tokyo conference. It is not known why Kanavarioti did not respond to the first letter, and it is unknown whether she responded to the second, as she had normally done in the past.

Even though Kanavarioti was not officially a member of the council of the IPPF before September 1954, nor was the HES their formal representative in Greece, they were treated as such. In July 1954, Houghton sent to Kanavarioti copies of two letters regarding two Greek gynaecologists, Dr. George P. Andritsakis and Dr. Angeliki Tsacona, who were interested in family planning. In the first case, the gynaecologist Andritsakis visited Houghton in London in April 1954, a little earlier than Kanavarioti, who visited her in May 1954. He asked for her guidance about family planning. Houghton advised him to meet Louros, because he was presumably an expert. However, Andritsakis reported to Houghton that Louros, although aware of the work of the FPA, he did not show any interest in family planning in Greece. According to Andritsakis, Louros found the idea for family planning promising, but inapplicable in Greece.[7] Houghton preferred to give the available information directly to Kanavarioti and let her handle the situation.

For Andritsakis, but not so much for Louros at this point, Greece needed family planning. He asked, therefore, the name and address of Kanavarioti and, if any, the details of the companies that sold contraceptives in Greece. As mentioned, however, Houghton forwarded his letter to Kanavarioti. It is not possible to deduce the reason why Louros disappointed Andritsakis but accepted Gamble's offer for contraceptives less than a year later, in February 1955. A possible suggestion would be that Louros was not officially the president of the HES before August 1954, so he had not organised its activities in Greece by July 1954, when Andritsakis contacted him.

As far as Tsacona was concerned, Houghton informed Kanavarioti that she was acquainted with Clarence Gamble. She was a gynaecologist, based on Thessaloniki, who had spent some time in the USA taking a special course in Gynaecology at the Free Hospital for Women, in Brookline, Massachusetts. The fact that she studied in the USA during the 1950s suggests that she came from a wealthy family. She planned to return to Thessaloniki in June 1954. Obviously, Tsacona was influenced by Gamble who was very active during this period in the USA. Tsacona sent him a letter claiming that birth control was indispensable in Greece due to the many large families. She argued that the country was overcrowded and contraception was deemed necessary. Therefore, she asked for supplies of diaphragms and jellies, in order to distribute them in Thessaloniki.[8] Tsacona was probably unaware of the illegality of distributing female contraceptives in Greece.

Tsacona was another example of a Greek physician who studied abroad, adopted new ideas regarding birth control and had the desire to spread this knowledge and practice in Greece. Moreover, Tsacona was one of the first and most popular female gynaecologists in Thessaloniki. Her alleged certainty, however, that the country was overpopulated was not entirely accurate. She probably had in mind the city centres of Athens and Thessaloniki, which were indeed overcrowded, but mostly by lone economic migrants from the villages, whereas during the 1950s most large Greek families continued to inhabit the countryside.[9]

Gamble sent her a letter in January 1955 to confirm that she needed diaphragms and jelly showing his willingness to supply her with contraceptive materials.[10] Their correspondence continued for a couple of months resulting in Tsacona's acceptance of his offer for spermicide jellies and diaphragms. Gamble immediately arranged the shipment but also prompted her to send him feedback of her experience with her patients. Most importantly he added: 'I hope you will find them useful for the poor people in Salonica', which suggests that the contraceptives were primarily destined for the low social class, in order to impede the creation of poor, large families.[11] However, their deal initially fell through due to strict customs and formalities. In May 1955, Tsacona explained to Gamble that there were two obstacles to getting the boxes with the contraceptives, kept by the customs authorities. Firstly, the tax was substantial; and secondly, she had to acquire a special permission from the Department of Hygiene in Thessaloniki. Tsacona claimed that she could not overcome these difficulties and she would send the supplies back with regret.[12] Ten days after her letter, Gamble responded by proposing the alternative of the 'sponge and salt method'. He would mark the boxes with the rubber as 'free gift', to facilitate the import from the customs. Gamble did not give up, and as will be examined next, he invented ways and means to send contraceptives to Greece, despite the strict customs regulations. Tsacona finally received the rubbers along with instructions on how to cut them in pieces for individual use, how to prepare the salt solution and some cards to record each patient's reaction and results. Gamble's ultimate aim was to gather information from all the countries he supplied with contraceptives both for his own research and international distribution. In his letter to Kanavarioti, he asked: 'Has the time yet come when it is possible to open a birth clinic for the poor people of Athens?'[13] Later, in a letter to Tsacona while negotiating the shipments he wrote: 'If you still feel that these materials [contraceptives] are needed for the poor people of Salonica [Thessaloniki], I will be glad to send you some without charge' repeating that the supplies were primarily destined to poor people, not for every woman who wanted to plan her family.[14] Gamble's objective was specifically to control the reproduction of poor people. The links between eugenics and poverty dated back in the beginning of the twentieth century (Brabrook 1910; Titmuss and Lafitte 1942). Obviously, both Gamble and the Greek gynaecologists shared the same goal, to limit the births of poor, large families. Along the lines with pre-war eugenics, it became evident that the control of reproduction of less advantaged people was also a goal of post-war eugenics. However, there was a substantial difference in methodology which was attributed to the change of sociopolitical constructions. Pre-war eugenicists dictated forced sterilisations of the poor or underprivileged, while post-war eugenicists recommended contraception and family planning education. Eventually, the motive remained the same and proved the continuity of eugenics ideology during the post-war period. Controlling the reproduction of certain groups of people justified the culture of selection entangled with eugenics throughout its historical course.

Furthermore, Gamble and his associates always sent patient cards and tabulation sheets to be filled up by the gynaecologists. Gamble asked Tsacona and Louros to record the progress of the patients who used the contraceptives in order to estimate their usefulness. Keeping these records was the prerequisite for supplying them with contraceptives. Therefore, Gamble reached the Greek gynaecologists on the one hand in order to control the fertility of the poor and on the other hand to measure the effectiveness of the available contraceptive techniques.

Due to the illegality of contraceptives, however, gynaecologists were very reluctant to advise on contraceptive methods and supply their patients with contraceptives. Added to this, there was the obstacle of importing them from abroad. When Gamble tried for the first time to ship contraceptives in Athens and Thessaloniki, it was not possible to pass the customs. The following times he sent them in boxes with the label 'for vaginal use' or 'medical samples'.[15] Therefore, the results from contraceptive use were not as quick and successful as Gamble would have desired. The progress was rather slow in Athens and almost non-existent in the countryside.

Both Andritsakis and Tsacona were obstetrician-gynaecologists who were interested in family planning but did not belong to the HES. This did not seem to be a problem for Gamble, although he developed closer relationships with the members of the HES. Furthermore, from the IPPF's point of view, the members of the HES—Kanavarioti in particular—were the first to be contacted in Greece for family planning matters. For instance, on 2 November 1954 Houghton informed Kanavarioti about someone who was travelling from Britain to Greece, to whom she had suggested contacting Kanavarioti and Andritsakis, assuming that Louros was not willing to get involved with contraceptives at that point.

Given that in September 1954 Kanavarioti became a member of the Governing Body of the IPPF, it was therefore reasonable for Houghton to get in touch with her concerning the visitor from Britain. Seemingly, the HES at this time had been the official contact of the IPPF in Greece; thus for every person seeking information for family planning in Greece, the IPPF did not hesitate to suggest a contact with the HES. Rotha Peers, for instance, also introduced people to Kanavarioti, such as her friend Mrs. Winter, who has a house near Athens to try and see you while she is over. Peers thought that Winter would be very interested in this work and she might know one or two people who would be helpful to Kanavarioti.[16] The remarkable ability of the IPPF to work worldwide cannot be divorced from the commitment of its members to their common cause.

Kanavarioti, on another occasion, responded to Tom O. Griessener from the IPPF office in New York about a request for contraceptives. The implied story was that two Greek people contacted Griessener asking about the availability of contraceptives in Greece, so he forwarded their letters to the HES. Obviously, the HES had already been actively preoccupied with birth control. At least that is deduced by Kanavarioti's response that she read the two Greek letters enclosed therein which ask for contraceptives and was pleased to inform him that the HES would contact with the writers and supply them with contraceptives.[17]

Judging from the short and confident answer, the distribution of contraceptives was common practice. The most significant detail is that Kanavarioti did not commit those people to Alexandra Maternity Hospital or any other clinic, but she assured him that the HES would contact the enquirers directly. This probably meant that the HES mediated between people seeking contraceptives and the clinic which distributed them. Otherwise the gynaecologists and members of the HES supplied contraceptives to their patients from their private practice.

2   Visits Abroad

The fact that Kanavarioti played an important role in the creation of the HES is beyond dispute. Unsurprisingly, she represented the HES abroad too. Her most significant visits were to Stockholm, London and Rome. As will be shown these visits strengthened the relationship of Greek eugenicists with international institutions. Kanavarioti's personal contact resulted in enjoying respect and admiration from her colleagues abroad even when she retired from her participation in the Greek eugenics activity.

2.1   Stockholm

Elise Ottesen-Jensen, one of the strongest supporters of the birth control movement and Sanger's successor in the presidency of the IPPF, was Swedish. Given that after the Second World War Sweden was one of the strongest states in Europe, Ottesen-Jensen organised a series of meetings in Stockholm, beginning with one held in 1946 (Blacker 1964). Kanavarioti visited Stockholm in 1953. The personal handwritten letter to William Vogt is of utmost importance for the history of the HES because in it Kanavarioti referred to this meeting in Stockholm. There she had the chance to meet Vogt, Sanger, Ferguson and Rama Rau. It seems that this was the first time that she met these people. Kanavarioti was jubilant, as expressed in her letter to Vogt.[18]

The IPPF held its annual conference in Stockholm in August of 1953 (Peers 1955), but the meeting to which Kanavarioti referred took place much earlier, because the letter to Vogt was sent on 10 March 1953. Supposedly a preliminary meeting took place prior to the official gathering. However, in a letter sent by Houghton on 27 April 1954, it was implied that Kanavarioti attended the official Stockholm conference. It was then that Houghton introduced the British Dr. Pyke to Kanavarioti. Unfortunately, the existing documents do not provide further information, so it remains unclear whether Kanavarioti attended both meetings in Stockholm in 1953. Notwithstanding this, it was important that she had the chance to meet these established population experts. The letter to Vogt included the information that Kanavarioti and Jiji Raue visited Egypt. Beforehand she had very interesting talks with Dr. Mantellos, President of the temporary Board of the Hellenic Eugenics Ass., Mr. Fylaktopoulos, Mr. Makris, Labour leader and many others.

As has been noted, Mantellos was the first president of the HES and President of the Athens Medical Association, and Fylaktopoulos was a psychologist, professor at Athens College and one of the leading members of the HES. Fotis Makris, on the other hand, was a very active Labour politician in Greece and one of the most important trade union leaders (Harry S. Truman Library and Museum). Considering this, it is unusual that he did not participate in the future activities of the HES. He was, however, noted by Kanavarioti, perhaps because of his popularity.

2.2   London

A significant step towards the development of the HES's international relationships was Kanavarioti's trip in Britain in May–June 1954. Houghton, as the Executive Secretary of the IPPF's office in London, corresponded with Kanavarioti to make all the necessary arrangements. In April 1954, Houghton sent a letter outlining the details of the trip. Interestingly, the letter was posted to an address in Oxford, which meant that Kanavarioti was already there. She returned to Greece on 15 June 1954.

She thus had ample time to visit experts and institutions including the North Kensington Marriage Welfare Centre, the FPA and the Islington Family Planning Clinic. Meetings with individuals included Mrs. Hobson from the Oxford Family Welfare Association; Mrs. Irene Heaton from Oxford Marriage Guidance Council; Dr. David Pyke of the Radcliffe Infirmary in Oxford, who she had already met in Stockholm in 1953; and Mrs. van Oss, who was Joint Treasurer of the FPA and associated with the Slough and District Married Women's Advisory Clinic. Furthermore, Kanavarioti had a meeting with Dr. Wheatherall, who was the Education Secretary of the British Social Biology Council. Houghton had previously sent her a copy of Wheatherall's paper on sex education in England which was presented at the Bombay Conference. The FPA organised the third International Conference on Planned Parenthood in Bombay, India, in November 1952. Kanavarioti was interested in learning about sex education in schools and Dr. Wheatherall was a specialist in this field. Not surprisingly, then, the HES would deal with this issue in a future conference. Most importantly, Kanavarioti met up with Carlos P. Blacker, the vice-president of the IPPF and secretary of the BES. The meeting was arranged for 12 May 1954 on Blacker's invitation.

Moreover Kanavarioti met Mrs. Cecily Mure, who was connected with the Walworth Women's Welfare Centre which was in turn affiliated with the FPA. Before their meeting, in December 1953, Houghton wrote to Kanavarioti on behalf of Mure regarding the latter's visit to Greece. The delegation of the IPPF's representatives intended to raise interest in family planning in Greece, a topic that also featured highly on the HES's agenda. Houghton sent a letter to introduce Mrs. Cecily Mure who has for many years been actively connected with the Walworth Women's Welfare Centre which was the first Women's Welfare Centre in Britain to give birth control advice in 1921 and was affiliated to the FPA. The organisation and layout of the Walworth Centre quickly became a model for other clinics. Moreover, Houghton mentioned that the methods of contraception taught by the doctors at Walworth have become standard practice throughout the country, and have been studied by doctors and other visitors from overseas. As the contact person between Kanavarioti and Mure, Houghton asked the former to assist the latter in her fieldwork in Greece and to facilitate meeting with Greek doctors and others who are interested in this field of work.

Yet Mure and Kanavarioti did not meet this time. In her letter to Kanavarioti, dated 14 February 1954, Mure explained that it would be difficult to arrange a meeting, because she would stay only for a couple of days. Nevertheless, Mure suggested a phone conversation on 18 February. Although it is not known whether this phone conversation took place, as mentioned before, Mure and Kanavarioti eventually met in Britain in May 1954.

During Kanavarioti's trip in Britain, Houghton additionally suggested attending the FPA's Conference of Branches and its annual meeting. Houghton motivated Kanavarioti to involve herself with family planning in general and associate with certain institutions in particular. However, Houghton admitted that she had probably suggested many more appointments than Kanavarioti would wish to keep, but acknowledged the opportunity of meeting a number of people in Britain who work in the family planning and marriage guidance movement. She assured Kanavarioti that none of them was obligatory to accept. In addition to the trip arrangements, two booklets regarding the work of family planning clinics in Britain, the *Clinic Handbook* and the *Family Planning: The Past and the Future,* were included in the letter. As Houghton explained, the booklets would give her an idea of what to expect to see at the clinics and of the history of the family planning movement. In her response, Kanavarioti happily agreed to participate in the activities that Houghton had proposed. Kanavarioti's interest in learning about family planning and keeping in close contact with these people was keen. The trip to Britain was an opportunity to associate with the IPPF's experts and visit family planning clinics.

At a more personal level, Houghton invited Kanavarioti to stay at her flat in London for some days during her absence, reflecting on the development of a close friendship and trust between Kanavarioti and Houghton. Although the content of this letter was informal, it had a letterhead with the IPPF's logo. Judging from the content of the letters before and after the trip, Kanavarioti was warmly welcomed, and the trip proved very successful and fruitful. Houghton's role was decisive for this positive outcome. Kanavarioti in return hosted Houghton at her house in Athens shortly after the end of the conference and meetings in Rome.

A personal relationship also developed with Dorothy Brush, to whom Kanavarioti mentioned Houghton's visit in Athens.[19] Dorothy Brush's daughter, Silvia, was married to a Greek man, so she regularly visited Greece. In addition, Kanavarioti's daughter, Leelia, lived in the USA and was a friend of Brush's daughter and her husband.[20] Kanavarioti was invited to attend their wedding in Greece as well. Just one day before

Kanavarioti's return to Greece (14 June 1954), Brush sent her a letter arranging to meet with her and some members of the HES. However, the meeting could not take place, because Kanavarioti was still in London. When they corresponded at the beginning of July 1954, Brush highlighted the fact that she did not try to meet anyone else from the HES without Kanavarioti's presence, because in her own words 'we all think of you as the leader'.[21]

Moreover, she commented that there was no obvious activity being carried out by the rest of the group. Indeed, the HES was then undergoing the first stage of its development, so Brush asked whether there were people, not necessarily members of the HES, interested in family planning to whom the journal *Around the World News on Population and Birth Control* could be sent.

Kanavarioti was the link between the HES and the IPPF because she was the one who contacted the foreign colleagues and met them in person, thus developing a profound relationship. Therefore, she enjoyed their trust and respect.

2.3   Rome

The culmination of the HES's effort to cultivate international relationships was Kanavarioti's attendance of the First World Population Conference in Rome from 31 August to 10 September 1954. The Economic and Social Council of the United Nations and the International Union for the Scientific Study of Population organised the First World Population Conference in Rome in 1954. The IPPF was represented by Elise Ottesen-Jensen and Dorothy Brush. From 1954 to 1994, the UN organised five world population conferences, indicating the importance of the topic at the time. These were held every decade during the twentieth century; thus in Rome in 1954, in Belgrade in 1964, in Bucharest in 1974, in Mexico City in 1984 and in Cairo in 1994.

A month before the World Population Conference in Rome, on 6 August 1954, the HES elected its new president and Executive Board. Louros succeeded Mantellos and became President, G. Pantazis, a professor of Biology, became Vice-President, Kanavarioti remained Secretary and S. Doxiadis became Treasurer. The remaining members of the Board were: physicians K. Konstantinidis, A. Mantellos and K. Saroglou; psychologists K. Katsaras and G. Fylaktopoulos; gynaecologist P. Panagiotou; and Professor of Paediatrics, K. Choremis.

Kanavarioti communicated the results of the elections to the IPPF. Houghton was delighted by the new composition of the Board and supported Louros' election. Along with her congratulations, Houghton sent Kanavarioti a formal invitation to attend the meetings of the Governing Body and Executive Committee of the IPPF.

Concurrently Kanavarioti received a letter from Vasilios Valaoras on 18 August. Apart from pioneering the study of biostatistics in Greece, Valaoras had made a career for himself in the UN. He was appointed a member of the Population Division of the UN and moved to New York in 1954. Before his departure to the USA, he had participated in the preliminary meetings of the HES and signed its statutes. As a member of the Population Division of the UN, he was going to attend the World Population Conference in Rome. As is indicated by Valaoras' letter, Kanavarioti had already informed him about the results of the election on 6 August 1954. Valaoras expressed his pleasure at the composition of the new Executive Board and its president, but most of all exalted Kanavarioti's work: 'One day our country will be grateful of the movement you started and the mastery of your work for this excellent beginning'.[22]

Valaoras also referred to his friend, van Vleck, and their discussions about Kanavarioti and the HES. He claimed that van Vleck's interest in the progress of the HES was equal to his own. He thus promised to persuade him to encourage the work of the HES. He informed her that both would be in Rome for the World Population Conference and that interesting matters regarding the work of the HES will be discussed there. Valaoras also referred to Houghton, showing that he was in contact with the IPPF. He concluded the letter with warm regards and congratulated her for the newly elected Board.

Valaoras was very supportive of the HES. On this occasion, he expressed his admiration for the HES's new Executive Board and his appreciation of its activities. Moreover, the fact that he asked Kanavarioti to attend the World Population Conference for the benefit of the work of the HES indicates his support of the eugenics movement in Greece, despite living in New York. Finally, it turned out that van Vleck was successful in convincing Kanavarioti to go to Rome. By his return in Greece, in early 1960s, Valaoras would rejoin the HES as a member of the Executive Board for many years.

The IPPF planned a series of business meetings after the end of the conference, where Kanavarioti was invited to attend as well. Indicative of their friendship was the fact that after Kanavarioti's trip in Britain, Houghton addressed her with her first name. Houghton informed Kanavarioti that the meetings were aimed at selecting new members; discussing policies and arranging the fifth International Conference, which was to be held in Tokyo in 1955. In this context, Houghton asked Kanavarioti to become a member of the IPPF's council, the Governing Body. She explained that the IPPF preferred Kanavarioti to other members of the HES, because they needed a working member, not a figurehead like Louros. She pointed out, however, that they needed Louros to accept an Honorary Associate membership in order to use his name to give prestige to their international organisation. Houghton enclosed a copy of the IPPF's first *Annual Report* and the *Constitution and Rules of the IPPF* to ease Kanavarioti's decision about becoming a member of the Governing Body.

Houghton and the rest of the IPPF members regarded Kanavarioti as the most active person in Greece, which was at the time true. Moreover, Kanavarioti was the most familiar, because she had already travelled to Sweden and Britain, where she met many of the IPPF's experts. Therefore, it was hardly surprising that she was the most successful candidate for that post. Although Houghton proposed an alternative candidate, Doxiadis, she made clear that their first preference was Kanavarioti. Trying to convince her, Houghton claimed that becoming a member of the IPPF's Council would help her local work. Furthermore, Houghton mentioned that van Vleck would also be in Rome and maybe in Greece afterwards.

Dr. Abraham Stone also verified Kanavarioti's presence in Rome in his letter, on 18 October 1954.[23] As can be deduced from the correspondence, Dr. Stone's visit and lecture in Athens was discussed during the meetings of the IPPF in Rome in September 1954. As Kanavarioti remarked in her letter to Stone, the members of the IPPF gave her much encouragement to continue the work in Greece. Stone was the director of the Margaret Sanger Research Bureau, deeply involved in family planning. Recently, Alison Bashford characterised him as the 'New York's contraceptive expert' (Bashford 2014). He was supposed to give a lecture on family planning at the Medical School of the University of Athens in January 1955.[24] Obviously, the IPPF was interested enlisting such a renowned expert as Abraham Stone to spread the word for family planning. However, Kanavarioti was obliged to cancel Stone's visit in the first place because it would coincide with the examination period at the university and sent him an apologising letter for cancelling his visit at the same time expressing her hope for a future one. The fear of a small audience resulting from the university examination period was one reason for the cancellation of Stone's lecture in Athens; the waiting for a better moment when the HES would be better established was another. In the responding letter, Stone expressed his willingness to visit Greece at another time and meet the members of the HES. He agreed, however, that the time was too short to prepare the lecture. Stone's persistent desire to visit Greece illustrates the IPPF's eagerness in conjunction with the Margaret Sanger's Research Bureau to include this country among their partners and to expand their international activities. As a result, Stone did not actually cancel the lecture, but merely postponed it. He ended his letter with the wish to successfully establish the Eugenics Society in Greece in order to disseminate information about the family planning programme.

3   Foreign Press

The most important step towards international recognition was the publication of the HES's establishment in the *Eugenics Review* in January 1955. Houghton explained to Kanavarioti that it was Blacker who wanted to include an announcement about the HES in the journal. The establishment of the HES was noted in the *Eugenics Review* as follows:

> Eugenicists in this country will be interested to hear that a eugenics society was founded in Greece in 1953. This is the Hellenic Eugenics Society, whose Secretary, Mrs. Maro Kanavarioti, was recently elected a Fellow of our Society. The newly-appointed Board, under the presidency of Dr. Louros Professor of Obstetrics and Gynecology at the University of Athens, held its first meeting on November 15th when it was decided to start activities in three main directions. These will include lectures to the general public, the first being given by Dr. Louros; the formation of a special committee to undertake the enlightenment of the Greek people through the medium of radio, publications, films, etc.; and a symposium of scientists to discuss current problems and carry out research [...] We feel sure that our readers will join with us in wishing every success to the new society. (Lane 1955)

The above publication represented the appreciation on behalf of the BES to its Greek equivalent. It was indeed rewarding for the HES to receive respect and support from the British eugenicists and be advertised in an as famous journal as the *Eugenics Review*.

Dorothy Brush, on the other hand, apart from the reference to the HES's successful lectures in the press,[25] she was eager to publish a more general article on the attitude of Greeks to birth control, abortion and sterilisation. She thus asked Kanavarioti to help her with the Christian Orthodox aspect by persuading the head of the Department of Theology at the University of Athens to write the Greek Orthodox viewpoint towards birth control, abortion and sterilisation. She had already published the religious views in the *Around the World News on Population and Birth Control* of Islam, Hinduism, Judaism and Buddhism. Moreover, Brush asked Kanavarioti's help to find the relevant Greek laws on similar issues in order to gather the legal texts of each country. Brush also thanked Kanavarioti for several new addresses that she provided her with, probably prospective receivers of the bulletin. She promised to find some educational films, such as the 'Biology of conception' and the 'Techniques of Contraception' which were designed predominantly for health professionals. However, she admitted that they had not published an equivalent paper for lay people.

Assuming that Kanavarioti was able to provide Brush with information about the Greek laws on reproduction issues; we can predict that it was less probable that she was able to inform her about the Orthodox Church's perspective on family planning. There was no official canon law regarding family planning. On reproductive issues, the Church was predominantly concerned with the matter of abortion, which was equated with homicide (Chatzinikolaou 2002; Mantzarides 2009; Vantsos 2009). Kanavarioti's response is not available though.

Interest in publishing something on the work of the HES was expressed by the American Eugenics Society (hereafter AES) too. Given that the British and American eugenics societies were directly related, the fact that the AES contacted the HES was not surprising. Frederick Henry Osborn (1903–1980) was one of the founding members of the AES in 1926 and the Secretary of the Galton Society in 1931. By 1946, Osborn had become president of the AES and radically transformed it into a more 'scientific' society. Following the mainstream tendency, he associated the AES with the population studies and birth control movement, which he strongly supported (Osborn 1952). Even in the American Eugenics Society Records, it is argued that Osborn's papers were instrumental in the shift in the American eugenics movement to a more scientific footing and into closer communion with population studies, and at the same time, they illuminate the link between population science and foreign and public policy in the post-war USA (American Eugenics Society Records). Remarkably, Osborn and John D. Rockefeller founded of the Population Council in 1952.

Osborn was made aware of the HES by his friend and colleague Pascal Whelpton and wrote favourably about it to Kanavarioti on 2 March 1954.[26] The purpose of the letter was to establish contact between the American and Hellenic eugenics societies and to introduce the journal *Eugenics Quarterly*, edited by the AES, to the Greeks. He admitted that the AES was at the time becoming more active and expanding its work. Osborn, indeed, expanded the work of the AES which remained active until 1972 when renamed Society for the Study of Social Biology (Gur-Arie).

Having in mind that Kanavarioti was Secretary of the HES, Osborn asked the editor of the *Eugenics Quarterly*, Mrs. Helen Hammons, to send to Kanavarioti a copy of the new journal to distribute it among Greek eugenicists. He also suggested including any forthcoming contribution from the HES in the journal. Remarkably, on the bottom of the letter was written that the programme of the society calls for continuing research and educational activity directed towards increasing the proportion of children born with better than average potential or intelligence and character and towards diminishing the burden of hereditary disabilities. It is significant that Osborn expressed his interest in the HES; although the ideological connection was obvious, the fact that Whelpton linked the two societies, even though he had visited Greece nearly two years earlier (December 1952), was very important.

As already mentioned, in October 1955 the fifth International Conference of Planned Parenthood took place in Tokyo. Houghton, who was preoccupied with its preparations, sent an informal invitation to Kanavarioti before the official invitations had been prepared. Houghton's comments are revealing of the financial difficulties facing this particular conference, which was going to take place in a region far removed from Europe and the USA. Houghton informed Kanavarioti about the financial difficulties the conference's location conferred. The American colleagues managed to raise money in order to send as many delegates to Tokyo as possible. For the European and Asian delegates, there was no funding; they should bear the cost on their own. According to Houghton, the Americans were motivated by personal interest in Japan. The population problem of the country should not be disregarded by them. The problem of the low birth rate in Japan was also a topic of discussion during the First World Population Conference in Rome. In a review of the conference, Peter Cox underlined the severity of the situation in Japan, where the birth rate dropped from thirty-three per thousand in 1949 to twenty-one per thousand in 1953 (Cox 1976). The author echoing the discussion during the conference attributed the low birth rate to the fact that the Japanese government passed the Eugenic Protection Law on the legalisation of abortions in 1948 (Cox 1976; Ogino 1996). The BES only supported Blacker's participation in the conference. Houghton was trying to raise money from the Rockefeller Foundation in Paris on behalf of Dr. Parker of the Medical Council to represent Britain, because he was in charge of the research on contraception. Moreover, Dr. Margaret Jackson from Britain would attend the conference by paying for herself.

In just one paragraph Houghton refers to the connections among the IPPF, the BES, the Rockefeller Foundation and the Medical Research Council in Britain. The most well-known collaboration was that between the Rockefeller Foundation and the Medical Research Council which had begun in 1923 (Mazumdar 1992). In this context, it was equally remarkable that she shared this information with Kanavarioti.

Regarding Greece, Houghton acknowledged the difficulty of raising so much money for the conference. However, she proposed that a delegate could stop off in Greece, an idea which was also expressed by Abraham Stone and Clarence Gamble. As one of IPPF's leading figures, Houghton knew Margaret Sanger personally and told Kanavarioti that she was going to meet her in the USA and travel together to Japan. Houghton probably followed her plan and made the trip, because their correspondence was interrupted for some months.

Rotha Peers, on the other hand, was about to prepare two reports on behalf of the European delegation to present to the conference in Tokyo. Regarding Greece, she had already prepared a short text on Louros' lecture 'Eugenics: An Appeal'.[27] Her desire to include this information about Greece revealed her enthusiasm for the lecture. This was a point where the two major events of 1955, Louros' lecture and the Tokyo conference, intersected. The success of the HES would be shared at an international level in Tokyo's conference.

The second report addressed issues of financial support in fields such as training, organisation, the foundation of clinics and propaganda. More precisely, Peers had to make a report of how the money of the IPPF could be better distributed across the European countries. In order to do so she demanded information and ideas about possible contributions. Furthermore, Peers suggested that an appeal for financial aid would be made by the IPPF to trusts and foundations, and if any help was given it would be to support a specific project for one or two years, rather than a grant over a longer period. Taking into consideration that Peers shared this information with Kanavarioti, the HES would be included in their plans for future financial aid.

Most importantly, the financial contribution of Joseph van Vleck and Dorothy Brush was a significant aspect of international collaboration. Funding is particularly mentioned because it was, in fact, the materialisation of external agents' confidence in the success of the newly founded Eugenics Society in Greece. According to Louros, Brush donated money to the HES to expand its activities. Although she was wealthy, it seems that the donation was an outcome of her acquaintance with the HES and particularly Kanavarioti. As extracted from the correspondence, Joseph van Vleck was Vasilios Valaoras' friend and colleague; thus probably it was Valaoras who introduced him to the eugenicist circles in Greece. Van Vleck initially participated in the activities of the HES as external funding agent and then, according to Louros, promised to cover forty-nine per cent of the HES' budget. Although there is no evidence that this indeed happened, there is no indication that it did not. Based on the assumption that he kept his promise, this offer was definitely a substantial aid to the HES. Van Vleck also contributed in other ways; he was present in Greece giving lectures on eugenics and family planning during the 1950s–1960s and communicated with both the Greek eugenicists and individuals outside of the country. Furthermore, it was revealed that important correspondence between Gamble and people from the HES was communicated with van Vleck too. Gamble's letter to Kanavarioti in 1953 also sheds light on the relationship between Joseph van Vleck and the Greek eugenicists because Gamble revealed to Kanavarioti that he got information for the activities of the HES by van Vleck.[28] Van Vleck's name appears in many other letters; for example, when in 1960, Gamble referred to van Vleck's visit in Greece in relation to family planning[29] and in Valaoras' letter to Kanavarioti in 1954 before the conference in Rome.[30] Van Vleck often visited Greece to give lectures, such as the lecture 'International Progress in the Field of Eugenics' at Alexandra Maternity Hospital in February 1958. His interest in Greece was keen and his assistance in the dissemination of eugenics and family planning was considerable.

4   CLARENCE J. GAMBLE

Undisputedly, the HES was internationally recognised in family planning circles. There was developed a mutual intercourse about the dissemination of birth control movement, primarily at the theoretical level. The contribution of Clarence Gamble added the practical dimension in this network. Clarence J. Gamble was an American physician and a millionaire, heir to the famous soap company Procter & Gamble. He was interested in the problem of overpopulation and considered birth control the only way to tackle it. His determination in conjunction with his wealth permitted him to travel the world and contribute towards the establishment of birth control clinics. Gamble also founded the New York Committee on Maternal Health and the Pathfinder Fund, which covered the cost of function of the birth control clinics and the salaries of his representatives.

At the beginning of the 1950s, Gamble became interested in Greece. He was aware of the absence of a birth control clinic in Athens and insisted on offering assistance. Gamble and fieldworkers associated with him, such as Edith Gates and Sarah Lewis, visited Greece many times, in particular during the 1950s and the 1960s. The purpose of their visits was to record and evaluate the situation regarding family planning in order to supply propaganda material and contraceptives. Their final goal was to establish a family planning programme in Athens that would include public education for 'baby-spacing' and the use of contraceptives. After every visit, a report was completed to be distributed among the members of the IPPF and Gamble's associations. There are seven reports dealing with Greece, from 1955 to 1961; four from Edith Gates, two from Gamble and one from Sarah Lewis. At the same time, and until 1964, Gates, Gamble and Lewis were in correspondence with members of the HES as well as with individuals associated with health institutions and gynaecologists in private practice in Athens and Thessaloniki. In the following years, Gamble supported family planning in Greece with the shipment of contraceptives and birth control information material.

The earliest of Gamble's letters to Kanavarioti is dated 23 December 1953.[31] It seems that they had already discussed the possibility to meet in Athens and Gamble informed Kanavarioti of the inconvenience of stopping in Greece on the way to India. However, he did not cancel the visit, only postponed it. More importantly, perhaps, the letter to Kanavarioti included Gamble's generous offer to support the popularisation of the HES, such as covering the cost of the propaganda printed material. In her response, Kanavarioti expressed her disappointment for the postponing of the visit, but she also hoped for another one in the future.[32] Kanavarioti added that she distributed one of Gamble's articles in Greek doctors in Athens. This was probably Gamble's most recent article: 'Human sterilisation and public understanding' published in *The Eugenics Review* in October 1953 (Gamble 1953). Its main argument in it was the possibility of influencing the government by public education. Gamble used facts and figures provided by the Human Betterment Leagues' activities across the USA to show that their campaign for sterilisation resulted in the sensitisation of the state officials and the passing of relevant laws. He justified his assumption that proper education (in fact manipulation) could lead to the acceptance of eugenics policies, such as sterilisation. While he did not specify it, he most probably discussed about voluntary sterilisation. In general, Gamble was a keen supporter of propaganda and public education; this is the reason why he wanted to help the HES with information materials and was so eager to send the journal, *Around the World News on Population and Birth Control*, to as many readers as possible.

The HES, however, decided to refuse Gamble's financial aid this time. The reason was the fact that the HES was not well established and was not prepared to accept this type of funding yet. The HES was also aware that before more formal relations were established with foreign organisations it needed official approval from the state. According to Kanavarioti, in a short period of time, the HES would be able to present its plans and projects to the general public. Then, Gamble's financial aid would be more substantial. Kanavarioti also underlined the moral encouragement that such an offer entailed.

The members of the HES were aware of their risky and novel task to familiarise Greek society with eugenics and family planning. Without support from abroad, it would be very difficult for the HES to grow and carry out its proposed activities. As was discussed in the previous chapter, there was little public activity and publicity around the HES prior to April 1954 when the official recognition from the Greek state was confirmed.

4.1   *The Distribution of Contraceptives in Greece*

Edith Gates visited Athens for the first time between 3 and 5 January 1955. Her first report discussed the illegal sale of contraceptives in

Greece, a matter which was the subject of many discussions among foreign organisations, such as the IPPF and Gamble's associations, and among some Greek gynaecologists who were interested in providing contraceptives to their patients. Copies of Gates' report were sent to the IPPF London, Sanger, Brush, Roots, Dick, van Vleck and Vogt.[33]

Dimitrios Poumpouras, obstetrician-gynaecologist and General Secretary of the Athens Society of Obstetrics, commented at the HES' meeting on 16 July 1953 on the difficulty and illegality of popularising contraceptives. Given that in Greece production or import of contraceptive devices or pharmaceutical preparation of contraceptives was illegal, any public education aiming at the diminishing of births or prevention of fertility would oppose the Greek law.[34]

A relevant legal document about reproduction, abortion and contraception was Article 305 of the Greek Penal Code, introduced on 1 January 1951 regarding the 'Advertisement of the means for the artificial termination of pregnancy' (Greek Penal Code 1951). It declared that: anyone who publicly with pamphlets, images or representations declared or advertised drugs or other subjects or ways by which he/she could provoke artificial termination of pregnancy or someone who offered his or someone else's services for that purpose would be punished with imprisonment for up to two years. However, it was not illegal to inform or educate about the interruption of pregnancy performed in public hospitals, if the information came from (1) a licensed physician; (2) a legal merchant of means for the artificial termination of pregnancy; or (3) a relevant publication such as a special medical or pharmaceutical journal. Article 305 did not explicitly prohibit the advertisement and trade of contraceptives, but methods and medication inducing abortion. Furthermore, this only referred to female, not male, contraception. This was probably the reason why Gates mentioned in her first report that: 'It is still absolutely illegal to do, be or give out contraceptives-the law still exists, though men may buy things at any news stand'.[35] She added that the time was not yet proper to send supplies, but mailing the journal *Around the World News of Population and Birth Control* was timely.

According to Gates, a more specific law prohibiting contraception was put forward by the National Hygiene Council (*Ανώτατο Υγειονομικό Συμβούλιο*) in 1957.[36] Louros, who was a member of the council, and other gynaecologists were, therefore, reluctant to promote the use of contraceptives in public. As was shown in Tsacona's case, it was very difficult to pass them through the Greek customs and to distribute them.

While the relevant legal text was not specific about the kind of contraceptives or their use, the majority of Greek gynaecologists were very cautious when dealing with the matter.

However, in 1955, the HES was the recipient of the first shipment of contraceptives in Greece, as revealed by Louros' letter to Gamble from February 1955.[37] Louros wanted to distribute the supplies at Alexandra Maternity Hospital, as he had hoped to start a campaign for family planning. To this effect, he wrote to Gamble about a discussion he had with Gates about the possibility to promote family planning and contraceptives in the outpatient department of the Alexandra Maternity Hospital. The acceptance of Gamble's offer was an important moment for the history of family planning in Greece. Until then, there was no active family planning programme or clinic devoted to it. Allegedly, the reason why Louros was in favour of family planning was primarily due to the fact that he was always against abortion. Furthermore, there was another reason relating to the issue of overpopulation in Greece. However, during the following years, Louros changed his attitude many times. Sometimes he was willing to distribute contraceptives to his patients; whereas in other instances, he showed no interest in family planning.

Betty U. Kibbee was one of Gamble's assistants who tried to find a solution to the illegal sale of contraceptives in Greece by contacting Mr. A. McIver, who was then a representative of the UN and High Commissioner for Refugees in Greece. Kibbee knew that McIver was transferred in Athens and tried to take advantage of his post in order to facilitate the importing of contraceptives.[38] Therefore she asked him if he could receive and distribute them in the country. In effect, Kibbee hoped that custom regulations would not apply to his position with the United Nations.

Notwithstanding legal obstacles, Gamble and his associates provided Greek gynaecologists and other health professionals with contraceptives by shipping them as 'medical supplies' or 'samples for vaginal use'. At that time the most popular female contraceptives were the sponge rubber, the diaphragm with spermicidal jelly and foam tablets (Kuyoh et al. 2002). While Gamble tried to send more diaphragms than foam tablets to the Greek gynaecologists, it turned out that the latter was preferable by both doctors and female users.

When Gamble finally visited Greece in February 1956, he met Louros, Panayiotou and Kanavarioti. He filled a report with regard to the situation about family planning in Greece. At the outset, Gamble's report referred to Louros and the difficulties he experienced in receiving supplies.[39] Gamble proposed the supply of foam tablets which were not marked as contraceptives, and could, therefore, be easily imported. Renaming contraceptives 'medical supplies' or 'patent medicine' proved to become the optimal way to avoid strict customs control. Soon it became the standard practice for the foreign suppliers of contraceptives to do this.

Following Gamble's instruction, in March 1956, Kibbee sent Louros three boxes each containing six diaphragms under the label 'patent medicine'.[40] About a month later, on 16 April 1956, Gamble also informed Louros that he had sent a large amount of the contraceptive jelly 'Metakol'. Although he feared problems with customs, he wrote Louros that an additional supply was ready to be sent. He would wait, however, until Louros was able to get them through customs without difficulty.[41] In thanking Gamble for the supplies, eighteen diaphragms and two hundred copies of Dickinson's book (Dickinson 1950), Louros also mentioned that he tried, in vain, to convince the state officials to allow the free import of contraceptives. However, he was optimistic and hoped that in the future they will succeed to convince them.

Meanwhile, the camouflaging of contraceptives as medicinal drugs continued. In a letter to Louros in 1957, Gamble talked about foam tablets, called 'Santronex', which could be sent from England. Again, they would not have been labelled as contraceptives, but as pharmaceuticals against vaginal germs. A few days later, Gamble confirmed to Louros that the Rendell Company, located in England, could ship foam tablets to Louros. Trying to encourage him to accept it, Gamble wrote that foam tablets were very effective in India and Pakistan, where he had the chance to test them.[42]

Kanavarioti was the person Gamble considered to be the most energetic in the HES and expressed his gratitude for her help while he was in Athens. Kanavarioti shared with Gamble her viewpoint that the public opinion regarding contraception was improving with reasonable speed. However, Gamble's plans to distribute foam tablets in the Greek villages seemed to be far-reaching. Gamble believed that the gynaecologists would limit their prescription to pathological cases, which was true at the time. According to Louros own words: 'the time has not yet come to say that contraceptive work is being done in Greece'. He explained to Gamble that the problem was political. Greek politicians believed that nothing should be done to discourage the multiplication of the nation because of 'the great number of Slavs at our back'.[43] Although Louros believed that the Greeks would be happier if they were half the number, most Greek politicians prohibited any means of birth limitation.

From mid-1961 to 1962, Edith Gates corresponded with a Greek import–export company, Chr. Nicolakis Company, based in Athens. The first contact was made with the owner of the company who was acquainted with Helen Stratigaki, working at the Aghia Sophia Children's Hospital in Athens. Nicolakis expressed his desire to establish a professional connection with Gates in order to import contraceptives and introduce this 'important health service to Greek women'.[44] Gates only responded several months later, in November 1961, by sending two consecutive letters.[45] She provided him with the relevant information regarding the manufacturing of foam tablets, in case he wanted to produce rather than import them. She informed the Rendell's Company about Nicolakis intention to start a business distributing contraceptives in Greece. Moreover, Gates referred to a Greek woman who probably had governmental connections and would facilitate the import of foam tablets. Her details, however, were not mentioned.

Meanwhile, Nicolakis had sent his request to the Rendells' company and also tried to reach an American company to provide him with the spermicidal cream-gel called 'Immolin'. Nicolakis wanted Gates' opinion about his new connections and about the product 'Immolin'. On Gamble's advice, Gates informed him that both gels and foam tablets were effective, but the foam tablets were cheaper. She was enthusiastic about Nicolakis' interest in contraceptives. Nicolakis responded with a thank-you letter on 3 February 1962, but after that the correspondence between them faltered.[46] Officially, the founding of a birth control clinic was only legally permitted in Greece in 1980, almost thirty years after Louros' first attempts to familiarise Greek women with female contraceptives.

4.2 Gamble's Delegates in Greece

Prior to Gamble's personal visit in Greece, Gates visited the country twice to assess the family planning situation. As already mentioned, Gates was one of Gamble's close associates, whom he financed to travel the world and popularise birth control. She had a particular interest in the Near East. There, she observed each country's activities for family planning in order to promote the establishment of birth control clinics and disseminate family planning techniques. In a letter to Kanavarioti on 20 December 1954, Gamble announced Gates' visit in Athens. Gates was already familiar with Greece, because she had worked there with the National Young Women's Christian Association. Gamble took the opportunity to repeat his offer for providing existing clinics with contraceptives, or to establish a new birth control clinic or provide an existing clinic with supplies 'for the poor people of Athens?'.[47]

On the same day, 20 December 1954, Houghton also contacted Kanavarioti to describe Gamble's activity, supposedly in preparation of Gates' visit in Athens. Houghton pointed out that Gamble was not a member of the Governing Body of the IPPF, but acted independently. She acknowledged his efforts internationally, mostly by visiting countries and funding activities related to family planning. However, his efforts were not always appreciated, according to Houghton, due to the 'unfortunate way in which he goes about the work'. Obviously, the IPPF did not always approve of Gamble's work, a fact which Houghton attributed to his attitude. She explained that Gamble's wealth made him a difficult person to collaborate with. Indeed, Gamble did not follow the code of practice of the IPPF or any other organisation; instead, he formed his own organisations based on his rules. Houghton aimed at informing Kanavarioti about Gamble and his delegates' behaviour before their visit to Athens.

Kanavarioti was the first person in Europe who met Gates. As she was in Greece, it was somehow geographically more convenient, because Gates often travelled to the Near East, which is closer to Athens than London. Houghton shared all the available information about Gates with Kanavarioti. Thus we know that Gates worked as field representative in the countries of the Near East, mostly Egypt and Turkey, disseminating ideas of birth control and founding family planning clinics funded by Gamble. Houghton informed Kanavarioti that none of the people in London had met Gates. However, they thought she had excellent qualifications and long experience in social work. She also made clear that Gates was not supported by the IPPF but the New York Committee on Maternal Health with which Gamble was associated. The emphasis put on the distinction between Gamble's work and that of the IPPF was prevalent throughout that letter. However, Houghton contacted Gates and advised her to meet Kanavarioti before attempting any other connection in Greece. Interestingly, Houghton made clear to Gamble that Gates should not attempt a meeting with Louros or Pantazis before contacting Kanavarioti and Gamble agreed. On the one hand, Houghton tried to keep Kanavarioti 'on the IPPF's side' but on the other hand she supported Gate's visit in Athens. Furthermore, Houghton made it clear to Gamble that Gates should not propagate family planning in Greece without first consulting Kanavarioti, as publicity might not be appropriate at this stage. Houghton's cautious words confirm that the time had not come for birth control propaganda in Greece by the time of Gates' visit in January 1955.

At the beginning of 1955, Rotha Peers replaced Nancy Raphael as Honorary Secretary of the IPPF for Near East, Africa and Europe. Consequently, she was interested in Gates' visits to these regions. On 23 February 1955, she sent a letter to Kanavarioti asking about her impression of Gates assuring her that anything she would say would be treated in the strictest confidence. Until then no one else from the IPPF had met up with Gates and Kanavarioti's opinion was considered to be the only source of information. In combination with Houghton's view that Gamble was not acting under the rules of the IPPF, the organisation wanted to know as much as possible about Gates' fieldwork.

In April 1955, Houghton repeated her inquiries about Gates, who had visited Greece in January 1955 and had already planned another visit for June 1955. She wanted to know Kanavarioti's opinion about the suitability of Gates for working for the IPPF. They wanted Gates to travel around in order to disseminate birth control and organise birth control associations at the local communities. The plan was to inspire and motivate the locals to organise a family planning movement in their region which would further receive supplies and training by Gamble's associations and the IPPF.

In asking Kanavarioti about Gates, Houghton indirectly expressed her feelings of trust and appreciation. Furthermore, Houghton was aware of Gamble's offer for contraceptives to the Alexandra Maternity Hospital in Athens. Alas, we do not have Kanavarioti's response to these letters. As a result her opinion of Gates is not known. However, one can assume that it was positive due to their excellent cooperation during Gates' visits. In January 1955, Gamble sent a letter to Kanavarioti, expressing his gratitude for her help during Gates' visit.[48]

In her first report on Greece, Gates described the Alexandra Maternity Hospital in Athens with obvious enthusiasm that the American funds aided Greece to obtain the most modern equipment. Alexandra Maternity Hospital was established in 1954 predominantly thanks to Louros. Among others, the hospital included a model School for Midwives and Nurses, the 'Queen Frederica' (*Σχολή Μαιών «Βασίλισσα Φρειδερίκη»*), where Louros taught. Furthermore, the first Centre for Prenatal Examination (*Μονάδα Προγεννητικού Ελέγχου*) in Greece was established there in June 1977. It was the first centre of its kind in Greece and the fifth worldwide. From its inception, Louros and members of the HES, such as Ioannis Danezis and Dionysios Kaskarelis, were directly involved in running this centre (Hellenic Eugenics Society 1978). One year later, in 1978, Louros aided the establishment of a Laboratory for Cell Genetics (*Εργαστήριο Κυτταρογενετικής*) for the diagnosis and prevention of congenital diseases and a Centre for Family Planning (*Κέντρο Οικογενειακού Προγραμματισμού*). It is, therefore, not a coincidence that many innovative methods and advances in gynaecology took place at the Alexandra Maternity Hospital. During Gates' visit Louros also presented the Sterility Unit (*Μονάδα Στειρότητας*) which he described as opposite to birth control, a description which caught Gates' attention. Louros and Alexandra Maternity Hospital were described by Gates in her report as follows:

> [Louros is] the proud director of the most perfect maternity hospital, the realization of a dream on which he has worked 17 years (and his father before him) and which he realises now through U.S. money. The equipment is complete from laundries and air conditioning to laboratories, every type of operating facility, research sections, sterility study, etc. This is to be the National Centre for all Child-Maternal Health work in Greece, connected with the smaller local centres, and the new experiment in Mobile Units starting in January, 1955 under UNICEF in Thessaly. He also teaches in University-doctors, midwives and training school for nurses in hospital.

She also characterised Louros as 'the leading doctor in the field', with a keen interest in family planning. This became evident in February 1956 when Louros asked for more contraceptives in a letter to Gamble, almost a year after his first acceptance of Gamble's offer. Louros must have received the first shipment of contraceptive diaphragms by April or May 1955 but used them much later probably by the end of the year or in early 1956.[49]

On the other hand, Louros made clear that the HES would focus more on education, rather than on running birth control clinics.

He openly insisted on presenting their work after the official publication of the HES statutes in February 1955. Kanavarioti and Pantazis outlined to Gates the content of the HES statutes and their plans. At the time of Gates' first visit to Greece, the HES counted forty members. The programme of public lectures was divided into three categories, each corresponding to three different target groups: (a) the general public (b) medical groups and (c) educational institutions (schools, universities, etc.). Twelve to fifteen lectures per year were scheduled on such subjects as genetics and heredity, demography and the history of eugenics.

Gates particularly appreciated Pantazis' work, because he was more practical than most members of the HES. Pantazis, who was Vice-President and the Chairman of the Educational Committee of the HES, claimed that education was the first step towards the implementation of a family planning service. Pantazis counted three major obstacles to overcome in Greece; the ignorance of the public; the Orthodox Church and the 'unpreparedness of doctors to help women, in fact their uncooperativeness, because they make money on abortions!' Abortion had been one of the major social-medical problems in Greece for half a century, contributing to the low birth rate and to deaths or injuries of women performing abortions in private practices. Therefore, Pantazis organised lectures for both lay and professional audiences to promote family planning. His plan included the establishment of a consultative Centre for Family Planning for the public and the introduction of sex education in schools. The latter was going to be carried out by doctors who could reliably give premarital advice to the youth. Furthermore, while Pantazis blamed the Orthodox Church for the difficulties of introducing family planning in Greece, Louros considered that the reasons were, in fact, political.

Gates' second visit to Greece in June 1955 lasted much longer and was more fruitful than the first one in January 1955. Again, her focus was on the activities of the HES, but she also approached people and organisations outside of it. She dealt mainly with women's associations, on which she reported details about their administration and activity. Moreover, she expanded the list of people to whom the journal *Around the World News on Population and Birth Control* would be circulated.

Already familiar with the leading people of the HES, Gates accepted their invitation to present her fieldwork in the Middle East during a meeting of the Executive Board of the HES. She reported that her presentation was well received. Louros, in turn, reported the successful organisation of two open lectures on eugenics. Regarding contraceptives, he hoped that these would soon be distributed in all the newly started maternal health programmes. In fact, he admitted that nothing had been done yet; Gates' comment: 'This is in the future!' shows her dissatisfaction with the situation. Given that Gamble had sent the supplies about two months before this meeting, she would probably have expected a more active plan. Gates mentioned that apart from Kanavarioti and Louros, also Pantazis, Konstantinides, Doxiadis and Katsaras attended the meeting.

During her second visit, Gates met Panayiotis Panayiotou, Associate Professor of Obstetrics and Gynaecology at the University of Athens and member of the Executive Board and Education Committee of the HES. Panayiotou was one of the gynaecologists who wanted to promote a simple form of contraceptives in order to be easier for doctors to apply and for individuals to accept. However, Gamble was not enthusiastic about his ideas, such as showing slides in cinemas or advertising on the public transport in Athens, and commented that 'this was more on eugenics rather than contraceptive lines'. Although Gamble thought that Panayiotou was not keen on providing his patients with contraceptives, he arranged that diaphragms and jelly were sent to him in October 1955. In December 1955, Panayiotou reported to Kibbee the difficulties he had experienced with customs, and it was only in January 1956 that he informed her that he had received them.

4.3   Contacts with Greek Associations and Women's Clubs

Gates summarised the problems of marriage and family planning in Greece in her second report in June 1955. First of all, she reported that the marriage ages between the sexes in Greece were very different than in Western Europe. On the one hand, men pursued their personal development and generally married between the age of thirty and thirty-five. On the other hand, women either began their sexual life very early, which resulted in many babies, or they chose to study first and then started a family between the age of twenty-three and twenty-five. Moreover, Gates mentioned the fact that priests got involved in the personal lives of the people, thus prohibiting contraception. As already mentioned, Pantazis acknowledged the Church as an obstacle in the use of contraceptives, whereas Louros did not.

Additionally, Gates emphasised the academic nature of the HES writing that Louros was 'still more concerned with the intellectual programs, not as aware of these down to earth problems of the poorer people'. According to Gates, the HES should have included more lay people in order to become more effective in tackling the everyday problems of marriage and procreation. She believed that the real family planning programme could be better applied in institutions such as the clinics of the PIKPA because they appealed to the general public.

4.3.1 The Patriotic Institution of Social Welfare and Awareness

Gates' first report brought to light PIKPA's importance in the birth control movement in Greece. Gates and others belonging to Gamble's foundation were interested in getting involved with the PIKPA, due to its large social network and its close relationship with Greek mothers. It was assumed that family planning guidance and supply of contraceptives would be easier through an already established network. The influence on women was also valued. Gates pointed out, however, that 'this must be tactfully handled as I understand the women in Greece each have their feelings of possession of 'their' society'. The report also described Kanavarioti's opinion that Tsaldaris was the leading woman in Greece who was also interested in family planning.

During her second visit in June 1955, Gates hoped to meet Tsaldaris and learn more about the function of the PIKPA. Kanavarioti mediated between Gates and some important people who otherwise could not have been contacted such as Tsaldaris. As Gates noted in her first report, PIKPA's network of clinics was ideal for family planning counselling. Tsaldaris was interested and in favour of introducing family planning advice, but she entrusted Dr. Saroglou, the Medical Director of the PIKPA, with the decision.[50]

4.3.2 The National Council of Greek Women

Kanavarioti also suggested Gates visit the National Council of Greek Women (NCGW). Gates was so impressed by its activity that she completed a separate section for the NCGW, attached to the main report. This was a union comprised of ninety women's societies, from Athens, Piraeus and other areas of the country and abroad. It was founded in 1908. It was acknowledged as a philanthropic institution and was under the patronage of Queen Frederica. It was also a member of the International Federation of Women's Clubs and in 1951 it organised the first international women's gathering in Athens, the Assembly of the International Council of Women. It was there that Tsaldaris gave a speech with the title 'The Child in Greece', in which she portrayed the history of child protection from antiquity to the twentieth century (Lina Tsaldaris Archive).

The NCGW's main activity was to help women overcome their problems, either personal or professional, and to defend their rights. Its fundamental principle was gender equality. There were fifteen different branches of action, one of which was concerned with health issues. Gates focused on it as a way to promote a eugenic programme. She estimated that organising a lecture on eugenics would be beneficial to their work. The NCGW published the magazine *Hellenia: The voice of Greek women* in English and a book series under the title *How to take care of your health*. *Hellenia* was published by the NCGW; the Lyceum Club of Greek Women; the Hellenic Association of University Women and the Hellenic Girls Guides Association.[51] Among numerous social causes, the NCGW also instituted a legal advice office, night schools, cinema shows for children, arts and crafts workshops and communal meals. Most importantly, it was very active in securing the repatriation of Greek children abducted by the Communists. For this purpose, it addressed appeals and protests to the UN and other international organisations, to mothers all over the world and to leading personalities. In addition, the NCGW succeeded in securing the right to vote for Greek women, in 1929 for municipal elections and in 1952 the right to be elected for parliamentary elections.

4.3.3 The Association of Female Greek Scientists

Pantazis arranged a meeting with Gates and Mrs. Katherine Papadopoulos, a member of the Executive Committee of the Association of Female Greek Scientists, another popular women's club in Athens. The AFGS published the journal *Halkyonides* (Ἀλκυονίδες), in which Dr. Popi Spelioti-Bazena, a gynaecologist and President of the AFGS, often discussed issues of eugenics, such as heredity, mortality, social instability, biological debilitation, hygiene and morbidity (Spelioti-Bazina 1950a, b, c, 1951a, b). The meeting with Papadopoulos was promising because she was interested in family planning and enthusiastic about organising public lectures on this subject. During the meeting, Papadopoulos expressed the AFGS's views on eugenics, more specifically they believed that eugenics was not only a science, but a social affair which concerns everyone. They endorsed the principal aim of eugenics which was the transmission of healthy traits to descendants and securing them the appropriate rearing environment. They also argued that prospective parents should meet certain standards, such as to be healthy, to be in stable financial situation and retain a psychological, corporal and spiritual harmony between them. Furthermore, they underlined that it was the duty of the state and the AFGS to organise the scientific study and the application of eugenic practices and to protect marriages from psychological and financial difficulties. Gates also noted that they claimed that sterilisation was necessary in some cases but not clarify the exact circumstances under which sterilisation was permitted or should be encouraged.

4.3.4 The Centre for Newborns 'The Mother' (Μητέρα)

Another institution reported by Gates was the Centre for Newborns 'The Mother', founded by Spyros Doxiadis and funded by Queen Frederica. It offered protection for unmarried mothers and orphans. While it was spacious, only a small percentage of the building was in use. It also hosted a School for Nurses, funded by the UN. It is suggested that its establishment resulted from the negative social perception and discrimination against unmarried mothers, who were helpless and marginalised in the 1950s. It was established in 1953, but became active in September 1955.

Prior to her third visit to Athens, Gates contacted Louros. His response was this time disheartening saying that he would be delighted to meet her again but he was very busy organising a conference. He also reported that there was not any significant progress in the matter of the distribution of contraceptives.[52] Paradoxically, when they met, Louros welcomed her warmly and was eager to report on developments in family planning in Greece. On the one hand, the National Hygiene Council opposed any work in family planning, claiming that there was no need for it, not even for poorer mothers. On the other hand, Louros received permission to run his own family planning clinic together with the sterility clinic at Alexandra Maternity Hospital, but only there.[53]

As far as the practicality of contraceptives is concerned, Louros reported to Gates that Greek women found the foam tablets easier than the diaphragms, which were most of the times unsuccessful. Louros expressed his desire to receive more foam tablets under the label 'samples for vaginal use', because the import of contraceptives was still illegal in Greece. Panayiotou shared Louros' view on the difficult use of the diaphragms. He argued that thousands of abortions were performed every year in Greece and suggested that foam tablets could be a solution to this problem. He also made a negative remark about another contraceptive method, the sponge with salt, which women did not like or trust.

After her return to Massachusetts, Gates became Director of the central offices of the Pathfinder Fund. Holding that position, she proposed a possible funding for the HES in a letter to Louros in October 1960.[54] She also informed Louros and Panayiotou about Mrs. Sarah Lewis, one of her colleagues, who was going to visit Athens in 1961.[55] By that time, Kanavarioti had retired and Marios Raphael became the new secretary of the HES. Moreover, Gates probably quit fieldwork and chose to offer her knowledge and experience through her new post at the office. New delegates, such as Sarah Lewis, succeeded her.

Louros agreed to welcome Lewis but informed Gates that the HES was no longer interested in family planning. Instead, the HES now focused on subjects of general interest. Louros repeated his position when he met Lewis some months later.[56]

Gamble was also aware of Lewis' trip to Athens and sent her a letter describing the situation. He recommended Kanavarioti, but she had already left for the USA. Based on previous remarks of Greek gynaecologists, Gamble advised Lewis to offer the foam tablets 'Santronex' produced by Rendell's Company or the 'Gynamin' produced by the Coates and Cooper Company. Somehow unexpectedly, he described Louros as 'an older man and because of his position in the Medical School, a conservative'.[57] This view was not shared by Gates, who praised Louros as 'a distinguished gynaecologist in the finest modern maternity hospital. He is most cordial and interested'.[58] Relying on her personal experience, Gates also suggested that Lewis meet Pantazis, Panayiotou and women's clubs such as the NCGW, the Association of Female Greek Scientists and the PIKPA.

Upon her arrival, Lewis contacted Louros, who immediately informed her that he was not concerned with birth control anymore. There is a paradox in Louros' thinking regarding birth control. As already mentioned, Louros was the first to introduce family planning in Greece in 1955, at his clinic at Alexandra Maternity Hospital when he received Gamble's contraceptives. By 1960, however, he had changed his mind. Louros justified his imbalanced attitude to Lewis with a series of arguments, such as the decline of the birth rate in Greece and the absence of birth control in the neighbouring countries. Tellingly, in 1956, Louros had argued exactly the opposite, when he told Gamble that he did not agree with the Greek politicians who thought that the population should increase in order to secure the borders of the country. Furthermore, in 1957 he asked for more contraceptives and propaganda material to be sent to Greece. In addition, in his 1960 article 'Fertility, sterility and overpopulation', Louros endorsed neo-Malthusianism and raised the danger of overpopulation (Louros 1960). He also recognised birth control as one of the most effective solutions to the problem of overpopulation. At the same time, however, he questioned some contraceptive techniques and wondered whether or not extended voluntary contraceptive methods may produce an involuntary sterility. Louros' argument was justified by his own observation of vaginal irritation after the use of foam tablets. Added to this, he argued that: 'It would be a grave national error for any nation to control its population while its neighbour's growth was not also limited'. In 1961, obviously because of the low birth rate, Louros claimed that: 'Greece could not be expected to use birth control when a vast frontier had to be guarded against so many adjoining countries, when those countries were not practicing birth control. The whole question hinges on the neighbours'. In almost all neighbouring countries, particularly Turkey and Egypt, the population was on the increase, a fact which caused insecurity in Greece. When Louros met Gamble a few months later, in April 1961, he voiced a moderate view that Greece could be better off with half as many inhabitants, but its safety would be jeopardised.

In 1962, Louros contacted the then General Secretary of the Eugenics Society in Britain, successor of Blacker, G.C.L. Bertram. Among other things, Bertram sent him a reprint of his article 'What are people for?' (Bertram 1961).[59] Louros' response was positive. He particularly appreciated Bertram's 'urge [...] for an international effort under the United Nations to produce a world development organisation so as to try to face the overpopulation explosion'.[60] In the aforementioned text, Bertram extolled contraception and worldwide population control. In his own words: 'Contraception is a vehicle for freedom and responsibility in the Western world. It is a blessing so far spread to only a small fraction of the world's population' and 'population limitation must indeed be brought about on the widest scale' (Bertram 1961).

Apparently, Louros may have supported birth control to tackle the world's overpopulation problem, but he thought it inapplicable to Greece, because it was not an overpopulated country.

The law prohibiting the distribution of contraceptives played an important role in his change of heart, because he tried many times to influence the National Hygiene Council and the Ministry of Health to change it, but unsuccessfully. However, he stated in his article (1960) that:

> [...] medicine can occasionally advise, but definitely cannot carry out a deliberate policy, especially where such a policy would have international repercussions. On the other hand, it is medicine's moral duty to work to improve the treatment of the individual sterile couple, although the problems and dangers of world overpopulation must be recognised and given immediate and serious study.

In addition, the HES's newsletter, published in 1962, presented the view of the US Ministry of Foreign Affairs (announcement no. 827) about population issues, represented by Mr. William Nanley. Nanley, as Louros explained, mentioned that it did not matter whether the population of India is five hundred or eight hundred million, but whether these people could be properly nurtured, dressed and accommodated. What was necessary was to develop our knowledge about population issues by advancing scientific, technological, social, political and economic research. The USA offered its expertise on population issues to other governments if requested. Moreover, Nanley mentioned that, even if it sounded absurd to Americans, birth control was not a central matter of discussion in many countries of the world. Given that Louros was responsible for the editing and distribution of this newsletter, it seems that he agreed with that opinion.

Two years later, in 1964 Sergios I. Mantalenakis, a gynaecologist and one of Louros' students, sent to him a letter to report his impressions from the second International Conference on Intrauterine Contraception in New York in October 1964. Louros' response disappointed Mantalenakis because he admitted that in Greece the issue of IUD could be discussed, but not applied as a regular practice because the use of contraceptives had been rejected by the National Hygiene Council long ago. Louros explained that the Greek population decreased; thus there was no interest in further diminishing it unless an international decision is made. Louros further added that he was very cautious with the use of contraceptives because he was aware of the harmful Grafenberg's device, which was then condemned by many gynaecologists (Mantalenakis 2004). Louros' letter confirms that by

1964 he had finally abandoned the promotion and use of contraceptives in Greece. He was, however, open to an internationally organised family planning movement. The biopolitical overtones experienced in Greece and the uncontrollable world population growth troubled Louros for decades, oscillating between one side and the other. Predominantly, Louros supported family planning, but he ended up conforming to the legal concept of encouraging births nonetheless. As Marius Turda has suggested, 'eugenicists—like other professionals—were frequently enveloped by their social and political existence, and often adhered to dominant social and political practices'; Louros aptly falls within this description (Turda 2010).

Therefore, during Lewis' visit, Louros limited the discussion to other medical issues apart from family planning, such as the drop in infant mortality rates and the raising of life expectancy rate. In addition, he highlighted the fact that there were too many doctors and not enough teachers, resulting in poor education levels in Greece. Moreover, Louros explained to Lewis that the HES was part of the Greek social welfare apparatus, therefore, when trained doctors from the HES contacted people at workplaces and offered medical advice; they showed anti-cancer films and organised public discussions. He also informed Lewis that contraception remained illegal except in cases where there was medical contraindication. When Lewis told him that she was more interested in maternal health of poor women, not birth control on a national basis, Louros happily put her in contact with the then President of the PIKPA, Mrs. Thalia Voyla. In a way, Louros transferred the debate about family planning from the HES to PIKPA.

Lewis took advantage of this connection and visited the PIKPA premises and discussed family planning issues with many people there. As a general impression, Lewis realised that apart from condoms, the majority of the Greek women were not aware of the other types of contraception. As a result, propaganda and education of midwives, social workers and teachers were deemed necessary. At the same time, the problem of numerous abortions persisted during the 1960s and Lewis wondered: 'how do we break the abortion racket among the hundreds of doctors who practice it?'[61]

Lewis was impressed by the PIKPA and its work. According to Voyla, the PIKPA received six to ten babies from Alexandra Maternity Hospital per week for adoption, but whenever was possible they persuaded mothers to keep their babies. Voyla showed Lewis around the kindergarten, took her to the children's rehabilitation centre in the Voula neighbourhood and to the 'Elliniko' children's home where she met Mrs. Mary Miller and Nitsa Th. Kalliga.

Dr. Tsakos was the administrator at another PIKPA's branch, in Penteli a suburb of Athens. Taking into consideration that he had studied hospital administration in the USA, Lewis believed that his foreign training would help. Tsakos was interested in family planning and asked for information materials and to receive the journal *Around the World News on Population and Birth Control*. Although optimistic, Lewis was cautious about how influential UNICEF and the WHO were with the PIKPA. She believed that these international organisations could discourage Tsakos from promoting family planning techniques in his institution.

Lewis also visited the Aghia Sophia Children's Hospital, where she met Mrs. Stella Megalou, Matron of the Nursing Service, who was in favour of family planning. Also present at the meeting was Mrs. Helen Stratigaki, Director of Education for nurses, who only knew of diaphragms, not the foam tablets or the sponges. Lewis wrote in her report that in a future visit Mrs. Sotiropoulou, Director of the Queen Frederica School of Nurses (hosted in Alexandra Maternity Hospital), should also be contacted.

Following Lewis' report on Greece and her personal meetings, Gates took the chance to approach more people and associations in Greece by mail. As in the past, she showed great interest in the PIKPA and sent a letter and relevant literature to Dr. Tsakos claiming that they were interested in popularising the idea of family planning to the poor and less educated people who allegedly did not understand the importance of the programme.[62]

Gates tried to convince him that the PIKPA with so many branches all over Greece could become the most strategic association to promote family planning as part of their regular prenatal and post-natal care, an integral part of the total Mother and Child Health programme. In conclusion, Gates expressed the Pathfinder Fund's interest in helping PIKPA in every possible way, but mostly regarding the supplies of contraceptives.

In a separate letter, having the same purpose as that for Tsakos, Gates approached Miss Elizabeth Papoutsidaki also working at the PIKPA's branch in Penteli and who also had lived in the USA. Gates sent her information on family planning and a few copies of Dickinson's book.

In addition, she sent some leaflets under the title 'Two simple methods'. Supposedly, the title referred to the foam tablets and the sponges that were easier to use than the diaphragms and simultaneously very effective. Gates informed her that the Pathfinder Fund was willing to send some samples of those simple methods to the PIKPA.[63]

Mrs. Helen Stratigaki, working at the Aghia Sophia Children's Hospital in Athens, received a letter and leaflets for family planning from Gates. The content of the letter was essentially the same as the one to Tsakos and Papoutsidaki. Gates recommended 'baby-spacing' and family planning programmes to alert the poorer, uneducated mothers who visit those centres. As implied in the letter, Stratigaki had asked Lewis to send samples of foam tablets, therefore, Gates sent her 'Santronex' foam tablets, produced by the Rendell's Company labelled as 'vaginal hygiene' and marked as 'medical samples for trial'.[64]

It seems that Gates tried to create a network in Greece, where the PIKPA, the Aghia Sophia Children's Hospital and the Queen Frederica School of Midwives would be joined under the leadership of Panayiotou. She suggested, therefore, contacting each other and uniting to promote this 'important health program to your mothers'. Gates relied on Panayiotou because he was the only one who really embraced family planning in Greece and could take action towards the distribution and use of contraceptives. Panayiotou was one of the few Greek gynaecologists who wanted to eliminate illegal abortions, but admitted that 'you cannot stop the abortionists', which indeed depicted the Greek reality. Lewis mentioned in her report that Greece at that time had four thousand illegitimate babies a year. In agreement with Louros, Panayiotou did not blame the Orthodox Church for the absence of contraception and differentiated it from the Catholic Church, who was strictly against the use of contraceptive methods. He asked for foam tablets labelled 'free samples for trial'. Apart from Panayiotou, Pantazis at Marika Iliadi Maternity Hospital received 'Santronex' foam tablets labelled as 'physicians samples'.

Gates sent Panayiotou a letter in which she referred to the people whom Lewis met and Gates corresponded with afterwards. Their target was to establish contacts with institutions where women most often visited to receive prenatal or post-natal advice. Therefore, Gates wrote: 'Maybe if the women took some positive action, the advice to mothers could be quietly integrated into the regular post-natal word'.[65] Meanwhile, she admitted that she had lost her faith in the work of the

HES: 'I begin to think the Eugenics Society isn't going to do anything, really, in family planning, is it?' Bearing in mind the discussion between Louros and Lewis, Gates was easily convinced that the HES was not going to continue the dissemination of birth control techniques. Instead, Gates turned her interest to institutions such as the PIKPA. However, she maintained contact with members of the HES, whom she trusted, such as Panayiotou, Pantazis and later Danezis.

Gamble visited again Greece from 19 to 25 April 1961 to participate in a conference organised by the Queen Frederica School of Midwives in Alexandra Maternity Hospital. In his presentation, Gamble discussed the problem of large families and provided information about simple contraceptive methods such as the foam rubber and the sponge. Following this visit, Gates sent relevant contraceptive supplies to the School of Midwives, again labelled 'for vaginal hygiene'. Gamble described the use of the tablets to Mrs. Sotiropoulou and offered some, but she did not want to accept them before getting permission from Louros.

While in Athens, Gamble also met Panayiotou who repeated his request for foam tablets. Panayiotou informed Gamble that he contacted the Greek Medical Association in order to change the law forbidding the importation and distribution of contraceptives and was optimistic. Panayiotou also tried to use the sponge and salt method, but his patients did not accept it and consequently rejected it.

Finally, Gamble met Louros and discussed family planning in Greece with him. Louros argued that the time was not appropriate for a change to the prohibitive law yet, and reiterated his position against sterilisation. Louros admitted that he tried to get along with the government and, therefore, used contraception only in few, extreme cases.[66]

Based on one of Gate's letters we learn about Mrs. James McEvoy's visit to Greece in the summer of 1962. In providing McEvoy with background information, Gates referred to the HES as 'a very cautious association which has only taken the eugenics approach and has been so fearful of the law of the land and the Church that they have not been willing to organise any family planning clinic—or even to use the term 'family planning'. Furthermore, she claimed that only Panayiotou, who was a leading gynaecologist and truly interested in the problem, actively supported family planning in Greece. He was keen on simple contraceptives which were more readily accepted by women. Gates also mentioned Kanavarioti, who although she had resigned three years ago remained the most active figure in the promotion of family planning and eugenics in Greece. Gamble repeated Gates' position on the situation in Greece, but also provided McEvoy with the information that Joseph van Vleck was in Athens to discuss birth control and proposed a meeting between the two.[67]

In the following years, the contact between Greek eugenicists and Gamble and his delegates faded. Not even Panayiotou was in contact with Gates or Gamble for some time. Gates sent him a letter in January 1963 and then another in 1964, probably without receiving any answer. Initially, Gates sent him a copy of the *Family Planning News* and a new pamphlet under the title: *Family Planning: A Challenge to Health Workers in Every Nation* in order to distribute it to social workers, nurses, midwives, etc. This was written in everyday language to be easier for more people to understand. The Pathfinder Fund was willing to send as many pamphlets as he wanted free of charge, but, as it seems by her next letter, Panayiotou did not respond. In the second letter, Gates informed Panayiotou about her meeting with Dr. Danezis at the IPPF conference in London. She also referred to the previous letter and demanded an answer about the distribution of the pamphlets and the situation regarding family planning. In addition, she informed him about a new intrauterine contraceptive method that Gamble also endorsed. He wanted to send samples to gynaecologists for trials in order to gather their reports and records.

During the same period, in September 1964, Gates contacted Ioannis Danezis following their meeting in London.[68] She complimented him on his work in educating doctors on family planning and the use of contraceptives. She asked if the Greek doctors had started to counsel parents about spacing their babies, something that Danezis already had been doing. Gates sent him copies of the above-named pamphlet *Family Planning: A Challenge to Health Workers in Every Nation* and the book *The Complete Book of Birth Control* and some samples of the new intrauterine contraceptive method. Moreover, Gates referred to Danezis' willingness to publish a leaflet for family planning in Greek and adapted to Greek customs. On behalf of the Pathfinder Fund, she assured him that 'We would be glad to make a financial contribution for this purpose'. Danezis was the treasurer of the HES from 1965 to 1967 and in 1974 was its president. He published regularly about family planning and his latest articles came out as recently as 2002 (Danezis 2002).

4.4   Propaganda

As mentioned before, Gamble paid particular attention to propaganda, public education and the transmission of contraceptive methods to physicians. He and his delegates in Greece disseminated family planning by personal meetings and lectures; by distributing the journal *Around the World News on Population and Birth Control* and by providing the gynaecologists with the then popular book: *Techniques of Contraception Control* by Dickinson. Pamphlets such as *Family Planning; A Challenge to Health Workers in Every Nation* were also distributed.

In her first report, Gates proposed the mailing of the *Around the World News on Population and Birth Control* to a list of people that she and Kanavarioti put together. Kanavarioti was characterised by Gates as 'still the lay leader, with a real sense of possession of 'her' organisation' and so she trusted her opinion. In every report, there was a section with a list of 'important contacts'. In her second report, Gates also included names of individuals who needed contraceptives, not only the journal. Birth control education escalated to birth control application, as was Gamble's main purpose. In a letter to Kanavarioti, Gamble estimated that they could send the journal to a hundred people in Greece, if Kanavarioti provided them with more names and addresses of people.[69] Gamble took the opportunity to request a list of people whom he wanted to receive the journal in his letter to Kanavarioti in January 1955.

Kanavarioti, in turn, sent him a list with the members of the Board of the HES, and other names including important scholars, such as George Alivizatos, Konstantinos Moutousis and Konstantinos Charitakis, all professors at the Medical School of Athens and supporters of eugenics; and Theodoros Vlissidis, Dean of the University of Athens. On 21 April 1955, Gamble informed Kanavarioti that he sent propaganda material and supplies according to her list. Later, Lewis filled another list of names whom to send the journal or samples of contraceptives or information materials. This time the enlisted people included either American or British nationals who lived in Greece, and some individuals working in women's and children's institutions.

Gamble's report in 1956 contained an overview of the work carried out by the HES, regarding public education. He mentioned that the open lectures continued and attendance was satisfactory. At the time when Louros was supportive of birth control, he suggested a more energetic plan which consisted of the publication of books and posters on eugenics in order to educate the patients of the Alexandra Maternity Hospital. He mentioned that the Alexandra Maternity Hospital coped with ten thousand cases per year, to which he had direct access to provide with family planning guidance. Gamble thought that the suggested budget of six thousand dollars for publications was very ambitious for a first attempt, but he agreed to send two hundred copies of Dickinson's book on contraception to be studied by doctors and students of medicine. Simultaneously, Dr. George Adamopoulos requested the same book which was distributed with the *Around the World News on Population and Birth Control* in November 1955.[70] Finally, Louros indeed received two hundred copies of the book in 1956, in order to distribute it to students at the Medical School and gynaecologists at the Alexandra Maternity Hospital. Panayiotou on the other hand expressed his desire to translate Dr. Dickinson's *Techniques of Contraception Control* into Greek.

Gamble and the Greek eugenicists favoured propaganda. During the 1950s and 1960s, the *Around the World News on Population and Birth Control* and Dickinson's book played also an important role in disseminating information about birth control, alongside the conferences and the open lectures regarding family planning and eugenics. Contraception and family planning were widely discussed in meetings and conferences regarding population problems, either international or local.

NOTES

1. Kanavarioti to Whelpton. 1953. *Nikolaos Louros Papers and Archive.*
2. Higgins, Margaret Louise (1879–1966), known as Margaret Sanger, was born in Corning, New York. She studied nursing at White Plains Hospital, but later she was mostly interested in sex education and women's health. She became a radical feminist and joined anarchist circles. In 1916, she founded the first birth control clinic in Brownsville, which at that time was considered illegal. As a result, she was imprisoned for 30 days. Some years afterwards, in 1923, she took advantage of a law which allowed physicians to found birth control clinics and opened one under the name 'Birth Control Clinical Research Bureau'. In 1929, she founded the 'National Committee on Federal Legislation for Birth Control', which favoured the dissemination and use of contraceptives. In 1939, she reshaped and renamed the 'Birth Control Clinical Research Bureau' as 'Birth Control Federation of America' and later, in 1942, as 'Planned Parenthood Federation of America'. During these years, she promoted birth control education, having the 'Birth Control International Information Centre'

as a cornerstone. In 1952, she succeeded in founding the IPPF 'the largest private international organisation devoted to the promotion of family planning'. See: The Margaret Sanger Papers Project, New York University. www.nyu.edu/projects/sanger/aboutms/about.html. Accessed July 2012.

3. Houghton to Kanavarioti. 1954. *Nikolaos Louros Archive.*
4. Marie Charlotte Carmichael Stopes (1880–1958) was the leading birth control and eugenics activist in Britain. See: University of Waterloo. Special Collections and archives. https://uwaterloo.ca/library/special-collections-archives/collections/stopes-marie; Hall, Ruth. 1978. *Marie Stopes: A biography.* London: Virago, Ltd; Rose, June. 1992. *Marie Stopes and the Sexual Revolution.* London: Faber and Faber.
5. Raphael to Kanavarioti. 1954. *Nikolaos Louros Archive.*
6. IPPF to Hellenic Eugenics Society. 1955. *Nikolaos Louros Archive.*
7. Houghton to Kanavarioti. 1954. *Nikolaos Louros Archive.*
8. Tsacona to Gamble. 1955. *Clarence Gamble Papers,* HMSc_23_77_1207.
9. See Hellenic Statistical Authority. Greek population censuses: 1951 and 1961. Research on workforce.
10. Gamble to Tsacona. 1955. *Clarence Gamble Papers,* HMSc_23_77_1207.
11. Ibidem.
12. Tsacona to Gamble. 1955. *Clarence Gamble Papers,* HMSc_23_77_1207.
13. Gamble to Kanavarioti. 1954. *Clarence Gamble Papers,* HMSc_23_77_1207.
14. Gamble to Tsacona. 1955. *Clarence Gamble Papers,* HMSc_23_77_1207.
15. Ibidem.
16. Peers to Kanavarioti. 1955. *Nikolaos Louros Archive.*
17. Kanavarioti to Griessener. 1956. *Nikolaos Louros Archive.*
18. Kanavarioti to Vogt. 1953. *Nikolaos Louros Archive.*
19. Kanavarioti to Brush. 1954. *Dorothy Hamilton Brush Papers.*
20. Brush to Kanavarioti. 1955. *Nikolaos Louros Archive.*
21. Brush to Kanavarioti. 1954. *Nikolaos Louros Archive.*
22. Valaoras to Kanavarioti. 1954. *Nikolaos Louros Archive.*
23. Stone to Kanavarioti. 1954. *Nikolaos Louros Archive.*
24. Kanavarioti to Stone. 1955. *Nikolaos Louros Archive.*
25. *Dorothy Hamilton Brush Papers.* Northampton, MA: Sophia Smith Collection, Smith College.
26. Osborn to Kanavarioti. 1954. *Nikolaos Louros Archive.*
27. Peers to Kanavarioti. 1955. *Nikolaos Louros Archive.*
28. Gamble to Kanavarioti. 1953. *Nikolaos Louros Archive.*
29. Gamble to Lewis. 1960. *Clarence Gamble Papers,* HMSc_23_77_1211.
30. Valaoras to Kanavarioti. 1954. *Nikolaos Louros Archive.*
31. Gamble to Kanavarioti. 1953. *Nikolaos Louros Archive.*
32. Kanavarioti to Gamble. 1954. *Nikolaos Louros Archive.*

33. Gates, Edith. 1955. Report to the National Committee on Maternal Health. Summary of Athens. *Clarence Gamble Papers*, HMSc_23_77_1208.
34. Hellenic Eugenics Society. 1953. Minutes of the second preliminary meeting. *Nikolaos Louros Papers and Archive*.
35. Gates, Edith. 1955. Report to the National Committee on Maternal Health. Summary of Athens. *Clarence Gamble Papers*, HMSc_23_77_1208.
36. Gates to McEvoy. 1962. *Clarence Gamble Papers*, HMSc_23_77_1215.
37. Louros to Gamble. 1955. *Nikolaos Louros Archive*.
38. Kibbee to McIver. 1956. *Clarence Gamble Papers*, HMSc_23_77_1209.
39. Gamble, Clarence J. 1956. Summary of Greece. *Clarence Gamble Papers*, HMSc_23_77_1209.
40. Kibbee to Louros. 1956. *Clarence Gamble Papers*, HMSc_23_77_1209.
41. Gamble to Louros. 1956. *Clarence Gamble Papers*, HMSc_23_77_1209.
42. Gamble to Louros. 1957. *Clarence Gamble Papers*, HMSc_23_77_1210.
43. Gamble, Clarence J. 1956. Summary of Greece. *Clarence Gamble Papers*, HMSc_23_77_1209.
44. Nicolakis to Gates. 1961. *Clarence Gamble Papers*, HMSc_23_77_1212.
45. Gates to Nikolakis. 1961. *Clarence Gamble Papers*, HMSc_23_77_1212.
46. Nicolakis to Gates. 1962. *Clarence Gamble Papers*, HMSc_23_77_1212.
47. Gamble to Kanavarioti. 1954. *Nikolaos Louros Archive*.
48. Gamble to Kanavarioti. 1955. *Nikolaos Louros Archive*.
49. Louros to Gamble. 1956. *Nikolaos Louros Archive*.
50. Gates, Edith. 1955. Report to the National Committee on Maternal Health. *Clarence Gamble Papers*, HMSc_23_77_1208.
51. On the first page of every issue, there was a paragraph explaining the reasoning behind the title of the journal: 'Among the many names of Goddess Athena, 'Hellenia' was the one by which she was known and revered by the inhabitants of Cyzicus in Asia Minor. Surrounded as they were by foreign peoples and influences, Hellenia symbolised, and kept alive in their hearts, the ideals and transitions of Hellenism' (Lina Tsaldari's Archive).
52. Louros to Gates. 1957. *Clarence Gamble Papers*, HMSc_23_77_1209.
53. Gates, Edith. 1957. Renewed Contacts in Greece. *Clarence Gamble Papers*, HMSc_23_77_1210.
54. Gates to Louros. 1960. *Clarence Gamble Papers*, HMSc_23_77_1211.
55. Gates to Panayiotou. 1960. *Clarence Gamble Papers*, HMSc_23_77_1211.
56. Lewis, Sarah. 1961. Report on Athens. Clarence Gamble Papers, HMSc_23_77_1213.
57. Gamble to Lewis. 1960. *Clarence Gamble Papers*, HMSc_23_77_1211.
58. Gates to Lewis. 1960. *Clarence Gamble Papers*, HMSc_23_77_1211.

59. Hellenic Eugenics Society: Bertram to Louros. 1962. *The Wellcome Library Collection*. http://wellcomelibrary.org/player/b16238096. Accessed 12 December 2014.
60. Hellenic Eugenics Society: Louros to Bertram. 1962. *The Wellcome Library Collection*. http://wellcomelibrary.org/player/b16238096. Accessed 12 December 2014.
61. Lewis, Sarah. 1961. Report on Athens. *Clarence Gamble Papers*, HMSc_23_77_1213.
62. Gates to Tsakos. 1961. *Clarence Gamble Papers*, HMSc_23_77_1212.
63. Gates to Papoutsidaki. 1961. *Clarence Gamble Papers*, HMSc_23_77_1212.
64. Gates to Stratigaki. 1961. *Clarence Gamble Papers*, HMSc_23_77_1212.
65. Gates to Panayiotou. 1961. *Clarence Gamble Papers*, HMSc_23_77_1212.
66. Gamble, Clarence. 1961. Report on Athens, Greece. *Clarence Gamble Papers*, HMSc_23_77_1214.
67. Gates to McEvoy. 1962. *Clarence Gamble Papers*, HMSc_23_77_1215.
68. Gates to Danezis. 1964. *Clarence Gamble Papers*, HMSc_23_77_1215.
69. Gamble to Kanavarioti. 1956. *Clarence Gamble Papers*, HMSc_23_77_1209.
70. Adamopoulos to Gamble. 1956. *Clarence Gamble Papers*, HMSc_23_77_1209.

References

Adamantidou, Triantafyllia, and Kiriaki Vantzeli. 2010. The History of the Alexandra Maternity Hospital. http://www.hosp-alexandra.gr/index.php?option=com_content&view=article&id=84&Itemid=67. Accessed 11 Jan 2012.

American Philosophical Society. American Eugenics Society Records. http://amphilsoc.org/mole/view?docId=ead/Mss.575.06.Am3-ead.xml. Accessed 16 July 2014.

Article 305. 1951. *Greek Penal Code. Around the World News on Population and Birth Control*, 35. 1955. Dorothy Hamilton Brush Papers.

Bashford, Alison. 2014. *Global Population: History, Geopolitics and Life on Earth*. New York: Columbia University Press.

Bertram, C.G.L. 1961. What Are People For? In *The Humanist Frame*, ed. Julian Huxley, 373–384. New York: Harper & Brothers Publishers.

Blacker, C.P. 1964. The International Planned Parenthood Federation. Aspects of Its History. Presented at the *Fourth Conference of the International Planned Parenthood Federation*, Western Hemisphere Region. San Juan, Puerto Rico 19–27 April.

Brabrook, E. 1910. Eugenics and Pauperism. *The Eugenics Review* 1 (4): 229–241.

Centre for Newborns 'The Mother' in http://www.kvmhtera.gr/index.php?detail_id=11. Accessed 12 May 2014.

Chatzinikolaou, Nicholaos. 2002. *Free from Genome: Orthodox Bioethical Approaches*. Athens: Stamoulis.

Clarence Gamble Papers, Series: III. Countries Correspondence and Records, 1927–1965, Box 77 (1207–1215) HMS c23. Harvard Medical Library, Francis A. Countway Library of Medicine, Boston, Massachusetts.

Cox, Peter. 1976. *Demography*. Cambridge: Cambridge University Press.

Danezis, J. 2002. The Contraceptive Pill in a Woman's Life. *Themata Meeftikis kai Gynecologias* 6 (4): 308–311.

Dickinson, Robert Latou. 1950. *Techniques of Contraception Control*. Baltimore: Williams & Wikins.

Dorothy Hamilton Brush Papers. Northampton, MA: Sophia Smith Collection, Smith College.

Doxiadis, Spyros. 1953. The Impact of British Nationalised Medicine to the Physician and the Patient. *Deltion Iatrikou Syllogou Athinon* 11 (1): 12–14.

Editor. 1959. Notes and Memoranda. *Eugenics Review* 51 (1): 44.

Foreign Relations of the United States. 1958–1960. Eastern Europe, Finland, Greece, Turkey 10, 2, 249. https://history.state.gov. Accessed 13 Nov 2013.

Gamble, Clarence J. 1953. Human Sterilisation and Public Understanding. *The Eugenics Review* 45 (3): 165–168.

Gardikas, Konstantinos. 1952. Medical Education in England. *Deltion Iatrikou Syllogou Athinon* 10 (10–12): 24–26.

Gur-Arie, Rachel. American Eugenics Society. 1926–1972. https://embryo.asu.edu. Accessed 16 July 2014.

Harry, S. Truman Library and Museum. www.trumanlibrary.org. Accessed 13 Nov 2013.

Hellenic Eugenics Society: Bertram to Louros. 1962a. *The Wellcome Library Collection*. http://wellcomelibrary.org/player/b16238096. Accessed 12 Dec 2014.

Hellenic Eugenics Society: Louros to Bertram. 1962b. *The Wellcome Library Collection*. http://wellcomelibrary.org/player/b16238096. Accessed 12 Dec 2014.

Hellenic Eugenics Society. 1978. *Public Discussions*. Athens: Parisianos.

Houghton, Vera. 1961. International Planned Parenthood Federation (I.P.P.F.). Its History and Influence: Part 1. *The Eugenics Review* 53 (3): 150.

Houghton, Vera. 1962. International Planned Parenthood Federation (I.P.P.F.). Its History and Influence: Part 2. *The Eugenics Review* 53 (4): 202–203.

Kanavarioti to Whelpton. 1953. *Nikolaos Louros Archive*.

Kuyoh, M.A., et al. 2002. *Sponge Versus Diaphragm for Contraception: Cochrane Database of Systematic Reviews*. http://www.ncbi.nlm.nih.gov/pubmedhealth/PMH0011866/. Accessed 16 Oct 2014.

Lane, K. 1955. Hellenic Eugenics Society. *The Eugenics Review* 46 (4): 198.

Lina Tsaldaris Archive, Konstantinos Karamanlis Foundation, Athens.

Louros, N.C. 1960. Fertility, Sterility and Overpopulation. *International Journal of Fertility* 5 (2): 171–173.

Mantalenakis, S. 2004. Ernst Gafenberg (1881–1957) & Gregory Pincus (1903–1967): The Pioneers of Intra-uterine and Hormonal Contraception. *Elliniki Gynecologia kai Meeftiki* 16 (2): 141–148.

Mantzarides, Georgios. 2009. *Christian Ethics.* Thessaloniki: Pournaras.

Mazumdar, Pauline. 1992. *Eugenics, Human Genetics and Human Failings: The Eugenics Society, Its Sources and Its Critics in Britain.* London: Routledge.

Nikolaos Louros Papers and Archive. N. Louros Foundation, Division of History of Medicine, Faculty of Medicine, University of Crete.

Ogino, Miho. 1996. Abortion, the Eugenic Protection Law and Women's Reproductive Rights in Japan. *Atlantis* 21 (1): 133–138.

Osborn, Frederick. 1952. Selective Process in the Differential Fertility of Family Stocks. *The American Naturalist* 86 (829): 203–211.

Peers, Rotha. 1955. The Fourth International Conference on Planned Parenthood. Report of the Proceedings, 17–22 August, 1953, Stockholm, Sweden. *The Eugenics Review* 46 (4): 255–256.

Rasidakis, N.K. 1953. The Psychiatric System in England. *Deltion Iatrikou Syllogou Athinon* 11 (9): 32–33.

Spelioti-Bazina, Popi. 1950a. The Twentieth-Century's Woman. *Halkyonides* 1 (1): 1–3.

Spelioti-Bazina, Popi. 1950b. Health as a Modern Issue. *Halkyonides* 1 (2): 5–7.

Spelioti-Bazina, Popi. 1950c. The Course of Sexual Instinct: Part 1. *Halkyonides* 1 (4): 3–6.

Spelioti-Bazina, Popi. 1951a. The Course of Sexual Instinct: Part 2. *Halkyonides* 1 (5): 3–7.

Spelioti-Bazina, Popi. 1951b. Sterility. *Halkyonides* 2 (7): 6–10.

The Eugenics Society. 1957. List of Fellows and Members. *Eugenics Review* 3: 16.

Titmuss, R.M., and F. Lafitte. 1942. Eugenics and Poverty. *The Eugenics Review* 33 (4): 106–112.

Turda, Marius. 2010. *Modernism and Eugenics.* London: Palgrave Macmillan.

Vantsos, Miltiadis. 2009. *The Ethical Consideration of Abortion.* Thessaloniki: K. Sfakianakis.

CHAPTER 5

Eugenic Concerns

1 Population Problems and Demography

The systematical observation of the Greek population's natural movement started in 1924 with the application of Law 2430/1920, which founded the General Statistical Service of Greece (Kotzamanis and Androulaki 2009). The Statistical Service introduced a new method for the registration of newborns. Every newborn was registered in a personal card which included the name, date and place of birth and other details. From 1928, the Statistical Service also gathered information from every other civil service that registered newborns because there had not been unified all registry services to one, as it is at present.

Few years after the proper establishment of the Statistical Service, Emmanuel Lampadarios and Vasilios Valaoras wrote an article on Greek population (Lampadarios and Valaoras 1939) as a response to the work of Dr. G. Banu *L' Hygiène de la Race* (1939).[1] Banu included the Greek population in the group of 'stable or ageing' populations but Lampadarios and Valaoras claimed that he had no accurate indications to defend his argument. Indeed, until the outbreak of the Second World War, Greece did not experience demographic decline, with the exception of periods of war (Kotzamanis 2000; Burnova and Garden 2014). Moreover, as discussed in the second chapter, the addition of approximately one and a half million refugees from Asia Minor to the mainland Greek population justified the increase in the total population.

© The Author(s) 2019
A. Barmpouti, *Post-War Eugenics, Reproductive
Choices and Population Policies in Greece, 1950s–1980s*,
https://doi.org/10.1007/978-3-030-03568-6_5

Lampadarios and Valaoras claimed that the Greek population was progressive until 1936 because the available means of research and measurement indicated an increase in birth rates and a decrease in mortality rates, particularly those of infant mortality. As a result, Banu's argument was unsupported by validated data. However, they admitted that only after the 1930s the Statistical Service produced and published accurate results.

Unfortunately, the Second World War and the German occupation were inhibiting factors for the further development of the Statistical Service. There were internal and external relocations which disorganised the administration. Until 1950, there were considerable efforts at the reorganisation of statistical services, although they were not successful. It was only in 1956 that the 'National Statistical Service of Greece' (*Εθνική Στατιστική Υπηρεσία της Ελλάδος*) replaced the first Statistics Service.

The most important Greek demographer, Vasilios Valaoras in the introduction of his work *Elements of Biometry and Statistics* (1943), defined and described statistics and biometry and their relations with eugenics. He established a connection between biostatistics and public hygiene by claiming that biostatistics was the only means of evaluating the results of public hygiene policies. Valaoras referenced Francis Galton's work on the research of genealogy. He claimed that Galton successfully applied statistics to research on heredity and praised his book *Natural Inheritance* (1889). In this book, Galton introduced, for the first time, methods of measuring the similarity among relatives in terms of 'bodily and spiritual dimensions' and personal habits. Furthermore, Valaoras expressed his agreement with Karl Pearson that Galton transformed the problems of evolution into problems of biometry (Pearson 1930). One of his beliefs was that there was no social equality, because lower classes and poor people were more exposed to diseases and death than the rich. As was discussed in previous chapters, Valaoras was also a member of the HES which dealt with population problems and demography.

The importance of demography and population studies was highlighted by the HES mainly at three conferences: in 1959 'The problem of overpopulation' (*Το Πρόβλημα του Υπερπληθυσμού*) (Hellenic Eugenics Society 1977); in 1974 'The problems of the elderly' (*Προβλήματα μεγάλων ηλικιών*) (Hellenic Eugenics Society 1974) and in 1975 'The reproduction problems of the Greek population' (*Προβλήματα Αναπαραγωγής του Ελληνικού Πληθυσμού*)

(Hellenic Eugenics Society 1975); but also on other occasions, such as in the conference held in 1971 'Environment and Survival' (*Περιβάλλον και Επιβίωση*) (Hellenic Eugenics Society 1978c). They discussed many aspects of demography: particularly the constant problems of sub—and over—population, infant mortality, urbanisation, differential fertility, low birth rates and the role of the state in these concerns. However, much earlier than these meetings, Louros addressed the problem of overpopulation and the need for family planning in his lecture 'Eugenics: an appeal' in 1955.

1.1   Overpopulation and Contraception

In 1958, a few years before the official announcement of the marketing of the contraceptive pill, the Greek magazine *Images* (*Εικόνες*) hosted a four-page article on it titled: 'A pill against Malthus' prophecy' with the interesting subtitle: 'Did science discover the best way for birth control?' The journalist portrayed the problem of overpopulation and presented the opinions of the Archbishop of Athens, Theokletos, and Professor of Obstetrics and Gynaecology Nikolaos Louros. It can be said that this article depicted the predisposition of the religious and academic-scientific points of view in Greece at the time. The Orthodox Church was not very much involved with population problems, while the academic community seemed more concerned. Interesting articles in the daily press appeared at the same time, such as the translation in Greek of an interview of Bernard Russell about overpopulation (MacKnight 1958). Although female contraceptives were not popular among Greek women in the 1950s, the fact that the invention of the oral contraceptive became a subject of discussion in a popular journal indicated that information about the overpopulation problem and contraception had already been a matter of concern in Greece.

Archbishop Theokletos needed no more than a few sentences to express the Orthodox Church's view on overpopulation. He claimed that the population growth did not pose any danger. He actually referred to a verse from the Bible about Divine Providence (Mt. 6: 22–33) saying that God could take care of every living being on Earth. According to the Archbishop, the matter of overpopulation did not pose any problems to the Orthodox Church. Although an encyclical was distributed by the Church in 1937 and another on abortion in 1968 opposing any means of contraception mainly because it was regarded as an intervention to the holy nature of the couple's intercourse which is likely to produce descendents, it did not gain much acceptance from the clergy (Stavropoulos 1981). In particular, priests who received the confessions of believers and were aware of everyday problems and anxieties disagreed with the encyclical (Stavropoulos 1977). The most popular response to family planning was abstinence which was non-invasive and harmless. However, even this kind of family planning method should be a common decision of the couple (Fanaras 2002). In this context, it should be noted that abortion as a contraceptive method has been unanimously rejected.

Louros on the other hand insisted on the view that the Greeks should response to the problems of the (poor) large families and sterility. According to him, both problems, although contradictory, could be sufficiently tackled with the study and application of a family planning strategy adapted to the best interests of the Greek race. He condemned strict birth control measures, such as sterilisation, but approved of regulation of births following the precise meaning of 'family planning'. Aware of the available means to plan one's family, Louros was positive in using them for the benefit of the family and the society.

In addition, Louros claimed that unless the state takes some serious measures regarding the problem of overpopulation, then academic discussions for population problems are pointless. He specifically recommended the parameters that the state should consider: the financial state of the citizen, the problem of housing and nutrition, the level of health and disease, subsidies, pensions, inadequate education, marriage and miscegenation, and finally, the pension age of workers. The essential point, however, was the prerequisite that this political movement against overpopulation should be implemented from the vantage point of eugenics, and not the standpoint of partisan interests. The goal of a family planning policy should be the improvement of the qualifications of the Greek race to the utmost limit, not to pursue political esteem. Once more, Louros showed that he had no political interests, either personal or on behalf of the HES. Louros aimed at convincing the state authorities that eugenics was important to the modernisation of the Greek society—thus deserved attention.

In this article, Louros brought to light the observation that rich people procreate less than the poor ones. While he referred to the relevant studies of Apostolos Doxiadis and Thrasyvoulos Vlisides about the disproportionate birth rates between high and low social classes, Louros openly questioned the absolute efficiency of rich children. He argued that growing up in a wealthy family could provide the child with the best of nutrition and education but it did not necessarily mean that the child would be successful or healthy, because genetic and social factors were equally important to a child's development. Louros did not support genetical determinism, and therefore, he often emphasised the environmental factor in human development during the HES public conferences.

On the financial level, Louros argued that the Greek economy, although improved after the wars, could not absorb the surplus of the Greek population in a few decades. In Malthusian terms, he claimed that if the Greek population decreased by half, the distribution of products would double. However, simultaneously such diminishment could result in military insecurity and the disappearance of the Greek race in the long run. Louros concluded his thoughts on overpopulation with Viscount Samuel who equated overpopulation with the H-bomb, but without finally expressed a concrete view about Greek demography.

The main body of the article was written by the journalist, who explained the situation with overpopulation and the distribution of goods on the planet along with the experiments and trials of 'the pill'. Initially, the journalist referred to the experiments in Puerto Rico, a country with a huge overpopulation problem and poverty. The first scientific indications showed that the pill was 100% successful and harmless; however, its possible side effects would be disclosed within five years. Presenting the studies of Malthus and Toynbee's opinion regarding overpopulation, the journalist was positive on the commercialisation of the first oral contraceptive. He also illustrated the statistics of countries with serious overpopulation problems, such as India, China, Japan, Egypt and other Middle Eastern countries. The article included historical facts about Gregory Pincus, who started doing research on the chemical constitution of the oral contraceptive in 1951 in Massachusetts and Dr. John Rock, an obstetrician and gynaecologist in Boston, who eventually collaborated with Pincus to improve the pill.

Remarkably, the author of the article discussed the possibility that population increase could lead to economic disequilibrium and war—a widely accepted opinion of the demographers of the 1920s–1930s (Bashford 2014). Therefore, he concluded the article with the hope that the poor and overpopulated countries receive the pill for free, when released onto the global market, in order to avoid the negative consequences of economic imbalance and war.

During the late 1950s, there was much interest in Greece in the problem of overpopulation as expressed in the press, such as the above-mentioned articles, and in conferences. For instance, the UN in cooperation with the Greek government organised the seminar on population problems of the Southern European countries, held in September 1958 in Athens. As such, the conference on overpopulation, organised by the HES in 1959, was timely.

Therefore, some months after the publication of Louros' views by the press, the HES organised one of the most important and popular conferences in its history with the same theme: 'The problem of overpopulation'. It was held in the hall of the Archaeological Society in Athens on 15 March 1959. This was a topic which concerned population experts worldwide since the beginning of the century and was inextricably linked with eugenics, biopolitics, geopolitics, emigration, unemployment and population control. The initiative for this conference was attributed to Louros, but Konstantinidis also insisted on discussing the importance of birth control.[2] The wide popularity of the conference, the attendance of Prince Peter and a large audience marked its success. The second newsletter of the HES devoted its largest part to this conference.[3] Moreover, the entire discussion was recorded by the National Radio Institution (*Εθνικό Ινστιτούτο Ραδιοφωνίας*).

After the conference, the newsletter of the HES included a report by P. Linardos, a journalist, who claimed that the conference on overpopulation was so successful for three important reasons. Firstly, the subject of overpopulation was timely and important; secondly, it was presented by a variety of experts, leading to a multidisciplinary approach, who were influential personalities, such as Louros, Goutos a sociologist, Goustis an economist, Merenditis, Colonel of the Hellenic Army, Panayiotou, Professor of Obstetrics and Gynaecology at the University of Athens, Pantazis, Professor of Zoology, and Svoronos, General Director of the Hellenic Statistics Service. As a result, the analysis of this conference illustrates the dominant views on overpopulation of eminent Greek scientists, scholars, health professionals and military officials of the given period.

In the beginning, Louros introduced the problem of overpopulation, outlining the classic Malthusian argument about the simultaneous multiplication of people and food shortage. He explained to the audience the theory that humans increased geometrically while the food supplies of the Earth produced arithmetically. Louros offered the most representative examples of geographies facing overpopulation, namely India, China and some African countries—always using the relevant statistics to justify his arguments. Louros posed some questions about overpopulation to stimulate the discussion, such as: 'Something has to be done about it, but what? Should we use birth control measures? This is not only a global, but also a Greek problem' (Hellenic Eugenics Society 1977).

Nikolaos Svoronos, as the General Director of the Hellenic Statistics Service, referred to the Greek population's movements since the nineteenth century, but focused on the period after 1930. Svoronos intentionally chose the period after the 1930s, because at that time there was no territorial growth as was the case in the period prior to 1930, with the exception of the annexation of the Dodecanese islands in 1947. Furthermore, the refugees who inhabited the country during the 1920s were integrated into the total population by 1930. As a result, it was more accurate to discuss Greek demography starting from the 1930s, after which there was no significant population change in the country. Svoronos informed the audience about the rise 24% in the Greek population during the period 1930–1956, while simultaneously there was a rise of 36% in the global population. Statistics show that Greece did not exceed the international level of population increase, resulting in the absence of an overpopulation concern. Svoronos portrayed the demographic situation of the period 1950–1956 when the Greek rate of population increase dropped to 8%. According to him, this was due to massive emigration not because of the drop of the birth rate. The Greek birth rate in 1950 was 19%, the same as the average European rate. In Asia and Africa, however, the rates were much higher up to 50%. Svoronos attributed these unequal rates to the use of contraceptives in the developed countries at the same time when in developing and underdeveloped countries people made limited or no use of contraceptive means. Regarding the average life expectancy rate, Svoronos presented the facts that in Europe the life expectancy from forty-three years for males and forty-seven for females increased in 1950 to sixty-five years for males and sixty-nine for females. In Greece, there were no official statistic tables for the average life expectancy, but Svoronos estimated it to be approximately sixty-five years. On the contrary, in the countries where the birth rate was very high, life expectancy was much lower than in developed countries. In India and some African countries, life expectancy did not exceed forty years. Svoronos claimed that the global population would be doubled by the year 2000 and would increase fourfold by 2043. He added that the food supplies resulting from the use of new forms of agriculture and other technological means of production could nurture ten to twelve billions of people. Svoronos was very cautious about the future because there were large populations facing malnutrition and poverty who might continue to be vulnerable to these dangers even if the science and technology of nutrition progressed during the following decades. Who would have access to these advances? Would technology advance according to the estimations? Would people be so educated as to use these advances to their benefit? These were some of the questions that Svoronos posed in order to highlight the importance of the issue of overpopulation in relation to malnutrition. He argued that only discussions at an international level could prevent overpopulation and its harmful consequences. However, he had a totally different opinion for Greece, because the birth rate had dropped to 16% in urban centres and 22% in villages, so there was no need to advocate for birth control. He concluded his presentation thus: 'Our national pride should not allow a nation such as the Greek, whose spirit had offered so much to the global culture and today represents only the 3% of the global population, to diminish its contribution to global culture in the future' (Hellenic Eugenics Society 1977).

Georgios Pantazis, Professor of Zoology and Vice-President of the HES, referred to overpopulation from a different perspective and through biology. He mentioned the process of natural selection which keeps nature in equilibrium. He argued that there was no possibility of overpopulation in flora and fauna due to the natural elimination of the unfit by the environmental conditions. Pantazis claimed that approximately the same process existed in aboriginal populations where infant mortality outweighed the population increase. Infant mortality and miscarriages were called 'natural checks' by population experts. The aim of birth control proponents was to replace these 'natural checks' of population with the free choice of contraception (Blacker 1954; Goodhart 1955; Connelly 2008; Bashford 2014). Aligned with the birth control movement, Pantazis embraced their theory of 'natural checks', as it appeared in aboriginal populations. 'The civilization', he suggested, 'which began by the white race and gradually spread to the rest of the world, created on one hand factors that impede nature's weapon, namely the natural selection, and on the other hand factors that facilitate the opposition to the natural selection' (Hellenic Eugenics Society 1977).

Pantazis then highlighted the fact that advances in medicine eliminated infant mortality and the spread of infectious diseases, which could be used as an excellent example of the suspension of natural selection. On the other hand, the elimination of births caused by voluntary birth control, not by some genetic factor, was also an example of interference with nature. Pantazis claimed the above-mentioned examples, while disturbing the natural balance—not only were desirable, but in some cases unavoidable.

According to Pantazis, birth control could be theoretically the most effective measure to avoid overpopulation that would be an inevitable outcome due to the opposition it posed to natural selection. He claimed that this was only an aspiration, practically non-applicable at the international level. According to Pantazis, the reason why birth control could not save the planet from overpopulation was that the civilised white race would apply this measure, whereas the other races would not. Therefore, according to Pantazis' overtly racist view, if the white race uses birth control techniques while the African and Asians do not, 'the white race will face the "yellow peril" and the world would be in racial imbalance' (Hellenic Eugenics Society 1977). As far as Greece was concerned, Pantazis believed that there was an unofficial control of births in urban centres where families with more than three children were rare. Obviously, Pantazis implied the popular practice of induced abortions in the cities as a method of family planning. In the countryside, there were large families because such birth control practices were not popular or because the ones used were not as effective as an abortion in private practice. However, he argued that it would not be wise to try to restrict the proliferation of Greeks because the population would automatically decline. At the time, Pantazis considered birth control in Greece undesirable and inappropriate, but necessary for overpopulated countries.

Panayiotis Panayiotou discussed the eugenics view, which concerned the quality of the population, not the quantity. Panayiotou argued that eugenics helped to understand the importance of the environment in human growth and development; therefore, human conditions, beneficial or harmful, were products of the interaction of both hereditary and environmental conditions. He unfolded his argument by discussing the importance of the social conditions in human development. The fact that socio-economic conditions, social justice and prosperity affect human development leads to the hypothesis that every law or institution could be a potential eugenic policy. Echoing Galton's perspective,

Panayiotou argued that the eugenicist should play the role of the 'natural selection' in society by replacing the rejection of the unfit (which happens in nature) with appropriate policy-making. Eugenics was applicable to hereditary diseases, such as haemophilia, incompatibility of the factor Rh of a married couple and epilepsy; also to socio-biological phenomena, such as marriage between relatives, adoption, artificial insemination; to premarital health certificate; and to general problems including the elimination of infant mortality. Moreover, Panayiotou supported the view that in order to achieve the optimal social organisation, instead of trying to genetically determine human, importance should be placed upon education, hygiene and intelligence. Sharing Louros view, Panayiotou rejected genetic determinism and emphasised eugenic intervention in managing the environmental factors of human development, such as education and nutrition.

Konstantinos Goustis, an economist, questioned Malthus' theory and finally rejected it by considering it a vague perspective that would not be universally applicable in every population and in every place of the planet—therefore useless. According to Goustis, Malthus discussed the relationship between the population problem and the sources of income and acknowledged a link; but this was by no means a solid theory to apply to the worldwide population. Goustis insisted that there was no general population problem because some areas of the world were overpopulated but others were not. There were different population tendencies which should be examined separately, and in their context. Regarding Greece, Goustis believed that the major socio-economic problem of the country was the high rate of unemployment. There was an immediate need to give Greeks the opportunity to work and be productive. Goustis' opinion was actually an attack to the supporters of the Malthusian theory and a rationalisation of the overpopulation panic of the time. He claimed that birth control was certainly not the solution; it was too strict a measure to impose. However, he supported family planning in the form of advice on the size of the family.

Michael Goutos, a sociologist who was interested in trying to answer the question whether birth control would be an effective measure in Greece, presented an overview of the latest official national censuses. He also mentioned that high birth rate does not necessarily mean that the population increases; it is always a matter of correlation among the number of births, deaths and emigration—a fact which was often highlighted by demographers to non-experts. Figures showed that the Greek population in 1950 was stable with a tendency towards ageing. As a result, Goutos was critical of birth control in Greece. As he put it: 'the survival of a nation is not only achieved by hygienic measures but mainly by high birth rate' (Hellenic Eugenics Society 1977). Regarding social policies, Goutos claimed that they favoured only the urban, working class, putting aside the rest of the members of the society. The first step should be to implement social policies at the national level. In agreement with Goustis, Goutos suggested facing the problem of unemployment and avoiding birth control.

Alexandros Merentitis, a Colonel in the Hellenic Army, undertook the responsibility of analysing the much-discussed issue of national defence in relation to birth control. Firstly, he drew a line between keeping the population stable and reducing it. If birth control did not result in the decline of the existent Greek population, then it would not hinder the defence of the country. Merenditis explained that the number of fighters was not as significant as many people believed. The possession of weapons of mass destruction was a far more important factor than a numerus army to achieve victory. Merentitis underlined the fact that a secure line of soldiers should only exist in the borders in case of a sudden outbreak of war. Therefore, he argued that birth control should be avoided in the provinces of Macedonia and Thrace where the borders of the country should be secured from a possible invasion from a neighbouring country. Merenditis' views were rather moderate; in fact, he contradicted the widespread argument that birth control should be avoided for the safety of the country which was employed by state authorities and at some point also by Louros. As he explained, the government should worry more about the armaments than the number of the soldiers.

Louros at the end of the conference introduced another view of birth control—its practical, medical application. As he put it, while there were many contraceptive methods, none of them was absolutely effective and on the other hand most of them could not be afforded by poor populations. Once again, it was highlighted that the concern of post-war eugenicists was the control of reproduction of the poor. Abortion was also a means of birth limitation, to which Louros was straightforwardly opposed. To strengthen his position, he referred to countries, such as the Scandinavian countries, Russia and Switzerland that permitted abortion for social reasons, but soon regretted it. Louros insisted on the equation of abortion with homicide, except when the mother's life was in danger. He questioned the idea of birth control per se and expressed his cautiousness for its practicality. As noted in the previous chapter, Louros expressed the same view in his letter to Mantalenakis in 1964.

Among numerous conferences of the HES, the conference on the problem of overpopulation was mostly published in the Greek press. Popular newspapers dealt with it and commented on the viewpoints of the presenters. In particular, the newspaper *Acropolis* (*Ακρόπολις*) published a series of eight articles on the conference. Their titles were impressive and eye-catching such as: 'The agonising problems produced by overpopulation. Is there enough space for the Greeks in Greece? A sensational discussion among seven top academics', 'Does overpopulation threaten Greece? If the civilised people apply birth control, the coloured will cover the Earth' and 'Does overpopulation threatens us? Birth control is not the number one problem of our country, but provision of labour to everybody'. Giorgos Koronaios, the author of the series of articles, portrayed the content of the discussion and the reactions and comments it provoked to the audience. He underlined the importance of the subject and the reputation of the presenters. The originality and audaciousness of the papers were also mentioned. It was also important that the first part of the articles was hosted on the first page of the newspaper. The first, and introductory article, included the editorial and Louros' keynote speech. This series was, in fact, the publication of the minutes of the conference in parts (Koronaios 1959a, b, c).

The presenters developed their arguments regarding overpopulation on the global scale, but also the population problem in Greece. While some supported birth control, others rejected it as inapplicable or inefficient. Contrary to the common Greek argument of the 'threat of the neighbours' used to justify aversion to birth control, Svoronos did not mention the possible military threats but focused on the safeguarding of 'national pride', while Merenditis put more emphasis on the kind of armaments, not the quantity of the soldiers. Pantazis on the other hand supported the global birth control movement only with international consensus. However, he considered birth control in Greece undesirable, due to population decline. Panayiotou, as expected, generally supported and insisted on eugenic policies, but not on the genetic level. Goustis and Goutos, who were non-medical professionals, added another dimension to the population problem, the high rates of unemployment. They argued that the most urgent problem of the Greek population at the time was that people did not have employment opportunities. The problem, of course, was proportionally aggravated by population increase.

1.2   Ecology

The growing concern for environmental disasters was the reason why the HES, in cooperation with the Archaeological Society, organised a conference on the environment. The presenters highlighted the dangers posed by environmental disasters and their reverberations for humanity, and the relations between the human behaviour and its surroundings. This conference was also concerned with population issues, although indirectly (Hellenic Eugenics Society 1978c).

Louros was again the discussion leader. In his keynote speech, he associated the environmental matters with the philosophical trend of positivism. He mentioned Johan Peter Frank, who coined the term and ideology of the 'hygienic police'; Christian Wilhelm Hufeland, who discussed for longevity; and Auguste Compte, who foresaw the problems caused by technological progress. In this way, he introduced the conference with a philosophical touch before permitting the presenters to express the practical view of the subject. Having already discussed the problem of overpopulation, Louros argued that the problem of overpopulation was crucial and agreed with Julian Huxley that each man would end up having one square metre to breathe. Overpopulation and urbanisation were indispensable parts of the discussion about the environment. However, Louros admitted that overpopulation was no longer a problem in Greece. Greece was an exception to the global overpopulation problem. On the contrary, under-population was the real problem of the country. He believed that the root of environmental disaster was the development and the uncontrolled expansion of industrialisation, which he equated with suicide. In the 1970s, Louros distanced himself from views supporting of birth control—once strongly advocated by him—due to the demographic decline of the Greek population during that period and his failed attempts to convince the Greek state authorities to implement policies supporting it.

Marios Raphael, a sociologist and General Secretary of the HES, referred to the science of ecology but focused on the struggle against disease and death. He argued that there was a continuous fight between humans and epidemics. On the one hand, the scientific and medical advances helped in the elimination of epidemics and the extension of the human lifespan. On the other hand, new health problems appeared, such as cancer, mental illnesses, allergies and others, which were caused primarily by the change in human's daily life due to technology; what we may call today 'lifestyle diseases'. Humans tried to control the environment, but ended up destroying it. The consequences of this behaviour were considerable and dangerous. Furthermore, man was isolated from the natural environment, losing contact with it. Most of the people lived in controlled artificial environments, where they did not see the daylight; did not feel the natural temperature; did not swim in the sea; but replaced all those ancient habits with new ones that fit to a man-made environment. Raphael argued that the way of living exerted influence on the environmental conditions and altered them; he argued that man needed more than just good health to survive; there were many dangers created by the lifestyle and the intellectual condition. He believed that George Orwell's dystopic novel *1984* (1949) was prophetic and it was highly probable to end up living in Orwellian controlled spaces under constant observation. The isolation in micro-societies could lead to the damaging of the environment and of the people living in it. Notwithstanding all these depressing thoughts, there was evidence that man survived by adapting to the environment. Therefore, Raphael concluded his presentation by claiming that it was highly possible for man to survive under any circumstances. He offered a positive view to the public in the hope that people would finally find a compromise between technological progress and environmental protection.

At the same time, Pantazis believed that since man was the only living being that knew about evolution, they had to try to command and control it for their own benefit. Pantazis, as a biologist, focused on the great importance of the role of the environment to human development. He argued that environmental factors had equal weight with the hereditary ones. Human organisms have mechanisms of fitting to the environment which allow them to survive irrespectively of the environmental changes, when those are not extremely intense or long-lasting. Pantazis used the word 'plasticity' to describe the Darwinian mechanism. Moreover, the environmental influence on someone's health was not inherited, because it did not influence the genes. It could induce anomalies or damages, but the person did not pass them on to his descendants. There were only a few types of environmental changes that affect the genes, such as some medicines, radiation and some chemical substances. According to Pantazis, the greatest environmental changes were artificial, not natural disasters. Man bore the responsibility of damaging the environment. Human choices, such as ignorance, indifference towards the environment and uncontrolled technological progress, brought disastrous results.

In agreement with Pantazis, Timos Valaes, a paediatrician and Director of the Institute for Child Health at the Aghia Sophia Children's Hospital in Athens, underlined the importance of the environmental factors to human development. He argued that every paediatrician deals with child development which is inextricably linked with the environment. He defined human development as a group of features such as aggrandisement, differentiation, growth, and spiritual and physical maturity. According to Valaes, people achieved fast growth and maturity by improving the environmental conditions. The environment and living conditions that they provided their children with allowed their genetic inheritance to be better manifested. During the preceding decades, people gained ten to twelve centimetres of height due to the technological progress and better living conditions. Although reluctant to admit that better environmental conditions resulted in a higher level of intelligence, he mentioned that there were studies that proved that under-nutrition was associated with low intelligence. Moreover, he referred to the side effects of urbanisation such as the damage of personal and social relationships and increased psychological stress. Drakoulidis, also member of the HES, had expressed in 1963 the same argument about the negative psychological repercussions of urbanisation in conjunction with overpopulation (Drakoulidis 1963). Valaes concluded his contribution with the reassurance that man was not deterministically a victim of blind evolution, but had the power to change his environment for his own benefit.

Ioannis Papaioannou, a Musicologist and Vice-President of the Institute for Child Health, focused on two main points regarding the environment: water supplies and nutrition. Even if water supplies seemed to be sufficient for the world's population, it was highly probable that serious problems of exhaustion of water supplies in the near future would be faced due to the increase in oceanic pollution. Papaioannou merely endorsed Malthus' theory about the gradual shortage of resources because of the growth of the population. He expressed his concerns about the fast increase in the population, which was much quicker than the increase in food production. However, in contradiction with Malthus' pessimism, he expressed his optimism that this problem could be solved by new food crops. He gave the example of wheat, which was planted in countries such as Mexico, India and Pakistan. This was very successful, because its production ended up to be more than expected and covered the needs of the countries in which it was planted.

Papaioannou expressed the ecological view in the matter of overpopulation, which was often underlined by population experts.

Added to Papaioannou's ecological approach, the dangers of air pollution were highlighted by Mariolopoulos, a former Dean of the University of Athens. The main point of reference was pollution from industry and the car exhausts. Frantzeskakis, a specialist in street traffic, added to Mariolopoulos' paper the urgency to confront the situation aiming at long-term outcomes. Resulting from overpopulation and technological progress, urbanisation and industrialisation were key factors for environmental disasters. It was in this context that birth control advocates strengthened their argument for the need of population management. Usually, this was expressed as an act of the well-educated population experts to the under-educated and poor people whose birth rate was uncontrollable.

Constantinos Doxiadis, brother of Spyros and an internationally renowned architect, summed up the environmental problems. First of all, he acknowledged a crisis in the relationship between man and the environment. He underlined the real dangers for man—particularly in an urban environment, such as the diseases caused by the intrusion of machines in human daily life. The uncontrollable use of any machine caused more harm to the people than good. Secondly, according to Doxiadis, too much information was another cause of problems. The wealth of information by television and radio made man dizzy and dangerous because of the lack of clear thinking. He also mentioned the damage that people caused to the monuments and the national heritage in general. Following on from this, pollution has expanded far from natural pollution to cultural pollution. Doxiadis believed that man could improve this situation by using technology in their favour. It would need to use the scientific advances with prudence, but also to be encouraged to make great changes to overpopulated urban areas.

An overall impression of the discussion on the environment was the fact that people should be watchful of the environmental disasters, because their implications could be catastrophic. While man exploited the environment and severely altered it by extensive use of technology, they could use technology to their benefit and save themselves and nature from disastrous outcomes. As far as birth control was concerned, although the participants did not exclusively refer to it, they emphasised the harmful impact of overpopulation to the environment. They all agreed that certain measures should be taken in order to reverse the current situation. On the one hand, human provoked the damage of the environment, and on the other hand, environmental disaster was depicted in human poor health. Human diseases often mirrored the damaging of nature. The harmful environmental conditions caused by human intervention to nature affected human body and mind and vice versa.

1.3   Population Ageing

Among various population problems, ageing was crucial because in conjunction with its demographic consequences, it stimulated important socio-economic changes. While the reduction in mortality was desirable, it was not advantageous if not accompanied by increased birth rate. Only in this case was there equilibrium in the quantity and quality of population. In Greece, from 1951 to 1971, the ageing of the population was both continuous and rapid.[4] This resulted in the decrease in morbidity and the increase in the average lifespan correspondingly. During that period, the socio-economic development affected the birth rates which gradually decreased. Simultaneously, emigration to the Western countries was massive.

The round-table discussion 'Problems of the elderly' was inspired by a conference at the Medical School in Athens in 1971. Dontas, an expert in gerontology in Greece and the chairman of the conference, appeared to have a cynical approach on this issue. He claimed that medical advances had a twofold impact: firstly, the lifespan was prolonged ten to fifteen years, and secondly, the lower classes benefited from the improvement of therapeutics, most notably preventive medicine. Furthermore, the elderly, who were 'less fit' for society, caused profound changes in the constitution of the population, because of their long lives. Not only was their care a financial burden for the rest of the society, but also they were isolated, both socially and psychologically, even when living with relatives. Dontas believed that the state and the society should adopt practical solutions to confront this problem. According to Dontas, health experts should reach a consensus on some striking issues: the definition of death; the time limitation of the living years of people in vegetate state; and the problem of euthanasia (Hellenic Eugenics Society 1965). He wondered if finally the price of individual longevity was the misery of the many—the rest of the society that cared for the elderly.

In general, even during the 1970s and 1980s, the majority of demographic publications referred to either low birth rate or the ageing population. These two problems were indeed the most alarming at the time. The scientific study of demography was neglected for a long period of time. On the contrary, 'less scientific' publications and articles multiplied. Kontzamanis, for instance, claimed that the discussions organised by the Eugenics Society about the above-mentioned problems and their consequences often highlighted a nationalistic approach to the present and future situation (Kotzamanis and Androulaki 2009). However, such an approach was at the time of the conferences reasonable, because every country cared for its own population and opted for its own improvement in quantity and quality. The devastating period during the first half of the century favoured nationalistic approaches on population which were widespread in the context of national reconstruction after the wars.

Drakatos, a demographer, expressed the popular belief that lower, poor classes multiply quicker than the upper classes. During the period between 1951 and 1971, the Greek middle class was the biggest portion of the population; its members had chosen to form small, nuclear families. As a result, Drakatos claimed that the low birth rate in Greece was due to the socio-economic development and that lower classes gave birth to more children than the middle and upper classes. Drakatos presented again the demographic situation in Greece which included the decrease in mortality rates and massive emigration to Western countries during the decade 1960–1970. Reflecting these changes in demographic patterns, Drakatos proposed a specific financial solution for the low birth rate and the nuclear families—approximately the same as that of Trichopoulos, but from a different point of view. Drakatos believed that the state should aid financially only the middle class, for example, the civil employees, who should get a 10% increase in their salary for every child until the third. He noted that the Greek population ought to increase not only in size but also in quality. According to Drakatos, the most effective pro-natalist policy would be to promote the creation of families having two or three children but from middle or upper classes, instead of the creation of large families having four or five children of the lower classes—a view shared by Spyros Doxiadis. If that plan worked, it would lead to a formation of a new category of people coming from middle and upper classes, who would not create any more social problems. Drakatos clearly stated that the adoption of this strategy would positively affect demographic evolution for the next ten to fifteen years.

Pepelasis presented a different perspective, focused on the problems of the workforce in relation to ageing. The ageing of the population was also financially multifaceted. On the one hand, the workforce, thus productivity diminished. In addition, there was the paradox that Greece 'imported' inexperienced and unqualified workers from abroad but simultaneously 'exported' Greek high-qualified professionals. On the other hand, financial help to a big part of the population burdened the state. Along with education and health, the financial burden of the elderly was the biggest economic problem of Greece in the 1970s.

From another point of view, state services sometimes substituted familial services. This was due to industrialisation. In rural areas, the family took care of its elder members. On the contrary, in urban areas, aged people relied on public services to survive. As Mousourou argued, the care of the elder members of the family was no longer the rule, but the exception. Even worse for the aged population, from 7% in 1956 it increased to 11% in 1971 but the public services remained inadequate for their care. Dimaki suggested the humanisation of the industrialised society as the optimal solution so as to achieve high quality of life for the elderly and smooth adaptation of the biologically younger elder in the family and society.

Furthermore, Dimaki referred to the psychological ageing of the population. Modern young people matured quicker than the past generations. Following Mead's outlook (Mead 1970), according to Dimaki, modern youth had equivalent experiences with older people of the past resulting in the undeniable fact that a modern teenager was the adult of the past. She argued that the terms 'ageing' and 'youth' had not absolute or static meaning; they were subject to socio-economic circumstances. She mentioned that the conflict of genealogies was inevitable when the elders managed society, because most of them occupied positions of authority.

Louros added to the discussion that Greece did not have fertility problems, but the excessive number of abortions resulted in the reduced birth rate. Moreover, urbanisation was another important factor of diminishing population. According to Louros, decentralisation against urbanisation and the wider use of contraceptives instead of abortion could help the country to revive (Barmpouti 2015b).

1.4   *Reproduction Problems*

Undoubtedly, issues regarding human reproduction were of primary importance to the Greek eugenicists not least because the majority of them were gynaecologists. As the core of eugenics rhetoric, reproduction intertwined with the broader concerns of the family, the social progress and the future of the Greek population. Therefore, fertility and sterility problems, abortion and the transmission of congenital diseases occupied the eugenics agenda of the HES.

The most renowned conference on the subject was 'The Reproduction problems of the Greek population' which took place on 20 March 1975 at the National Research Institute in Athens. By that time Louros was succeeded, firstly by Spyros Doxiadis (1973) and then by Ioannis Danezis (1974). Louros resigned as President of the HES because he was appointed Minister of Education in July 1974 and due to his old age. In 1974, he was seventy-six years old. However, he remained as Honorary President. Following the advance and popularity of genetics, the Hellenic Eugenics Society was then renamed as the 'Hellenic Eugenics and Human Genetics Society'. The conference was initially organised in 1974 in the context of, and as part of, the UN's celebrations of the 'World Population Year'. While there was a Greek delegation in the World Population Conference in Bucharest in August 1974, the political restlessness that prevailed in Greece had repercussions in the academic life and the conference in Greece was postponed (UN 1974).

In brief, Turkey invaded Cyprus in July 1974 and the Greek government of the military Junta was criticised for its poor strategy. Shortly after that first conflict, the dictatorship unable to confront the situation gave its authority to politicians. At that time, the new 'emergency government' under Konstantinos Karamanlis undertook the governance of the country. Unfortunately, neither the dictators nor the politicians managed to confront the sudden Turkish invasion which was repeated three weeks after the first operations. In November 1974, there were the first elections after the dictatorship in which the New Democracy, a political party again led by Konstantinos Karamanlis, won and gradually returned political and social stability in Greece. As a result, the conference was postponed for March 1975.

The discussion was coordinated by Danezis and the participants were: S. Doxiadis; Mrs. D. Milonaki, an economist; A. Pepelasis, a manager of the Agricultural Bank of Greece; N. Polyzos, a demographer and economist; D. Trichopoulos, an Associate Professor of Hygiene and Epidemiology; and D. Tsaousis, a Lecturer in Sociology at the Panteion University of Athens.

Firstly, Danezis emphasised that the purpose of the HES was to bring to light the world and national population problems, not to offer concrete solutions. According to the organisers, the ultimate goal was to disseminate knowledge about the problematic nature of subjects associated with population tendencies into the public arena. Danezis also referred to the World Population Conference in Bucharest (1974), which followed the World Population Conference in Rome (1954) and the World Population Conference in Belgrade (1965). What made the one in Bucharest unique was the fact that the delegates represented their governments, not an academic institution. Demography was inextricably linked with politics, a fact which was overtly shown during the conference (Finkle and Crane 1975). The population problem was addressed at the political level, a fact which made the signing of a common plan of action very difficult. However, a consensus was achieved by the majority of the participating countries. Some of the proposed actions were to promote the education and information of the general public on population and fertility problems; to take measures about the distribution of population in each country; and to improve the study of demography and family planning. The main aim was to promote health programs and social policies. The ultimate goal was the improvement of the quality of life (Mauldin 1974).

Valaoras was one of the four people who represented Greece in the World Population Conference in Bucharest. The leader of the Greek delegation was Andreas Kokkevis, Minister of Social Services.[5] Two works of Valaoras were distributed among the delegates of the conference. Those were the *Protein-Calorie Deficiency and Child Health* and the *Urban-Rural Population Dynamics of Greece, 1950–1965*. The presence of a delegation showed that the Greek state was concerned about population problems and demography. Added to this, Valaoras' work was highly appreciated and respected (Valaoras 1976).

Danezis focused on two outcomes of the conference: the fact that population was an important factor for the development of a country, and that gender equality in family matters was essential. Furthermore, each government was responsible for its population policies and reproductive problems. Emigration, urbanisation, poverty, energy supplies and education were also discussed as intrinsic aspects of the population problem as a whole. On a personal level, each couple should be free to decide whether and when to procreate and be responsible for baby spacing. As for the family planning institutions, which Danezis was very familiar with, these should be incorporated into the general health programs of each country so as to make citizens aware of family planning strategies and techniques. In the conclusion of his keynote speech, Danezis highlighted the insufficiency of the Greek demographic statistics and the lack of a demographic policy. 'Anarchy of reproductive forces' was his exact description of the Greek population problem (Hellenic Eugenics Society 1975).

More accurately, this was a round-table discussion, in the form of dialogue among the participants, not the typical presentation of individual papers. The discussion began with Trichopoulos' contribution, answering Danezis' question about the factors that shaped demography. Trichopoulos referred to the three major aspects of demography: reproduction; mortality and emigration. The outcome of their interdependence and intertwining depicted the population tendency of a country. More importantly, Trichopoulos analysed the situation in Greece. By the 1970s, the mortality rates had been decreased much below the world average. In particular, infant mortality, which was the most critical, had been adequately decreased too. There was, however, room for improvement. Low reproductive rates were the most alarming population problem of Greece during that period. Since the 1960s, there had been recorded a rise of nuclear families and at the same time an increase in the marriage age. It was that period just after the Greek women gained their right to be elected in parliamentary elections (1952) when their emancipation gradually occurred. Having access to higher education and professional development, the founding of a family was postponed to a later age by the modern Greek woman. As a result, the reproductive years became fewer and the predominant family model was the nuclear family. Trichopoulos was optimistic though because the trend of getting married at a young age, from twenty to twenty-five years, revived in the 1970s. This shift automatically meant that there were more chances to have large families. Moreover, positive was the fact that emigration rates gradually decreased when immigrants from the 1960s began to return in the 1970s.

Doxiadis, on the other hand, focused more on infant mortality. He agreed with Trichopoulos that there was improvement in infant mortality rates, due to medical advances, but he added that the inappropriate socio-economic circumstances should not be overlooked in the persistence of the problem. While medical progress and technology improved both maternal and child health, the lack of hygienic living conditions and proper education of the mother were factors which hindered the good health of the newborn. Often infant deaths occurred after familial negligence. According to Doxiadis, the number of the members of a family was crucial for child development. Based on the results of a research carried out in England, he argued that children who were descendants of large families (more than two children) did not manage well at school. He thus proposed that the ideal family model was that of two or maximum three children. According to Doxiadis, more attention should be paid to the increase in the children in nuclear families than to the multiplication of large families. This was a realistic and achievable solution, if equilibrium between quantity and quality was to be reached. Furthermore, Trichopoulos' opinion on financial aid for large families fitted neatly into this way of thinking. As he claimed, he had already discussed it with Louros and reached a consensus that the state should cut financial aid for families of three or four children because parents were tempted by the money and gave birth to children without having the means to raise and educate them properly. As a result, the number of illiterate and under-educated people was growing. As stated previously, illiteracy could lead to unwanted conditions of living.

In addition, Nikolaos Polyzos, the founder of the Greek Society for Demographic Studies, agreed that the poor were most vulnerable to disease and death. He attributed child and infant mortality to illiteracy and outdated baby nursing knowledge of mothers, particularly in rural Greece. Polyzos argued that illiteracy rates were commensurate with infant mortality rates. Therefore, the rate of infant mortality showed the cultural level of a country. The same idea continued to prevail in population studies. Infant and child mortality was also attributed to the lack of hygienic conditions of the lower classes (Valaoras 1950; Papavassiliou 1954). The living conditions and the environment where a child was born and raised were crucial. Again, it was claimed that the popularisation of hygiene precautions and childcare methods was imperative. Danezis, on the other hand, stretched the issue of the lack of prenatal care and medical observation of pregnant women. It was repeated that women in rural areas were prone to miscarriages and infant mortality. Moreover, in 1970, statistics showed that 82% of the total deliveries in Greece took place at the large maternity hospitals of urban areas.

Pregnant women living in the countryside, in their last month of pregnancy, moved to the big cities to deliver their baby, which was also a dangerous procedure for the health of the newborn. The lack of state and individual prenatal and post-natal care was clearly illustrated by the statistics.

Trichopoulos presented three factors playing the most important role to the diminishing number of births. First of all was, of course, the high rate of abortions, for some the eternal reproductive problem of Greece. Trichopoulos pointed out that secondary sterility added to the harm of the abortion per se. An abortion could increase the possibilities of sterility four times more than other causes. Secondly, the postponing of marriage affected reproduction rates, because in traditional Greek society, childbearing before marriage was a social taboo. Therefore, the combination of late marriage and absence of births before marriage resulted in fewer children. Thirdly, demographic research indicated that the Greeks preferred to have two children, on average (Valaoras et al. 1965). Trichopoulos insisted that the ways to achieve this number of children were contraceptives and abortion. This assertion provoked Danezis' reaction who argued that contraceptives were neither used at large nor suspended reproduction; instead, they helped couples to better plan their family and baby spacing. Danezis also argued that contraceptives were the antidote to abortion, that family planning advice helped raise the educational level of women—and the rest of the family—in health and reproduction issues and that control of reproduction should be the right of every woman. Danezis highlighted the fact that less than 1% of Greek women used contraceptives for the control of reproduction. As a result, the claim that contraceptives contribute to the decrease in birth rates was groundless.

Mylonaki presented the economic aspect of population dynamics. She claimed that in the short term the domestic and national economy might benefit from the low birth rate. In the long run though, low fertility would diminish the number of workers and reduce the level of a country's productivity. All depended on the government's population choices and policies.

In contrast, Pepelasis argued that there was no concrete evidence that low fertility provoked low productivity; these were only hypotheses. He referred to people's high physical and professional mobility, which influenced the fertility rates as well. Although Trichopoulos insisted that reproductive rates in the rural areas were very high, Pepelasis argued that the children of the villagers emigrated during their reproductive age, so the statistics were not realistic and the demographic problem of Greece was much more serious. Polyzos, as a demographer, insisted that statistic figures depicted the reality which was gloomy for Greece due to the diminishing of the number of children, emigration and the ageing of population (gerontogrowth) leading to degeneration.

Tsaousis shared Pepelasis' opinion about mobility, both geographic and social, with regard to the preferences in the family size. Moreover, he underlined the change in women's social image, having lost the label of 'reproductive machine'. Conjugal relationships were also ameliorated resulting in effective decision-making for family size from both parents.

At the international level, the participants mentioned the global population problem which was reflected in the high rate of births in underdeveloped or developing countries in contrast to the adverse rates in developed, mostly Western, countries. Although on the global scale births should be decreased due to overpopulation, in Greece the opposite should be the target. Louros—as a member of the audience this time—referred to the critical geographical position of Greece which demanded a robust army. As expressed before, the neighbouring countries continued to threaten the national integrity of the country. As a result, the global movement against overpopulation was at odds with the population problem in Greece. As Polyzos argued, the impeding of low fertility was not 'national selfishness' (nationalism) but the right of the Greeks to survive.

Population problems ranging from high density to desertification, from obesity to starvation, from robustness to epidemics and from over-productivity to under-productivity were discussed at large by demographers, sociologists, physicians and other population experts mostly with regard to the international scale. When discussed in the national context, however, the discussion focused on the nucleus of population, the family. Preoccupation with issues of the institution of family is indispensable to the wider population's concerns.

2   FEMALE EMANCIPATION AND FAMILY

The change of the regime in 1974, when the Government of the National Unity succeeded the military Junta, had repercussions on the legal framework of the country. The need to change the constitution set out by the Junta in 1968 was urgent. Therefore, in 1975, the democratic government adopted a new constitution to replace the former. One of the groundbreaking changes was the addition of Paragraph 2 of Article 4 which declared that: 'The Greek men and women are equal to the law and have equal rights and obligations' (Official Government Gazette 1975). The new paragraph of the constitution stipulated that the Greek men and women had the same rights. This simple sentence provoked a series of reactions on many grounds.

Following the legally established equality of the sexes, a series of discussions and meetings of experts took place in order to incorporate equality of the sexes into the entire Greek legal framework and society. A committee under the supervision of Andreas Gazis, a Professor at the Law School of the University of Athens, was responsible for changing and integrating the new family law into the former one. The committee examined the implications of the establishment of equal rights for both sexes, as reflected in the family life. They were responsible for adapting this major sociopolitical change, the equality of man and woman, into the family law of the Civil Code. Most importantly, equality of rights dictated the eventual collapse of the patriarchic model. For example, children's nurture and education were then both parents' responsibility. Added to this, there had been important changes in the matters of abortion, adultery and dowry. Moreover, the situation of single mothers was then legally supported.

As Gazis argued, the challenge of his committee was to replace father's authority with parental care. The essential meaning of this replacement was that the care of the children became obligatory for both parents. In the past, the authoritarian role of the father did not leave room for the mother in the decision-making of the family. The new family law regarded the mother as equal to the father. Both had to be in agreement with matters concerning the child. If they did not, then the law would protect the child. This was exactly the purpose of the new legal framework, to protect the child from a possible conflict between its parents (Hellenic Eugenics Society 1979). Other matters of concern were divorce and the function of single-parent families.

Gazis and Michalis Stathopoulos, a member of Gazis' committee, participated in the conferences of the HES. This was another example showing on the one hand the importance of the participants in the HES's activities and on the other hand, the connection between them and the Greek state. In particular during this period, members of the HES, such as Louros and Doxiadis, were also members or former members of the government.

In fact, equality of the sexes and the changes it provoked to the institution of marriage were effectively implemented with the passing of new laws at the beginning of the next decade. Among the significant legal innovations was the equation of civil with religious marriage in terms of legality in 1982 (Official Government Gazette 1982), and when the equality of the sexes was fully incorporated in the legal texts of the Civil Code, the Commercial Law and the Code of Civil Procedure in 1983 (Official Government Gazette 1983). In the same context, another law contained the cancellation of the previously compulsory law for the premarital health certificate and the legalisation of the family planning advice in 1980 (Official Government Gazette 1980). This law permitted family planning advice in public clinics and maternity hospitals along with the establishment of special units for family planning in ten regions of the country. A few years later, abortion was also fully legalised in 1986 (Official Government Gazette 1986; Ziegler 2008).

Establishing equal rights for men and women mirrored the Greek social reality. As elsewhere in Europe, during the world wars, Greek women also participated in the warfare either as heroines or as victims while safeguarding the survival of the country. Although female participation in the wars was largely neglected in historiography, it was an indisputable fact (Wingfield and Bucur 2006). Thus, it might be the devastating experience of the war that strengthened their personality and eventually led them to claim their rights after the end of it. In fact, women began to enjoy more respect in the male-dominant society since late nineteenth century when they were allowed to enter universities in 1890 (Varika 1987). Although there were few women who desired and achieved such an accomplishment, they paved the way for the next generations. Gradually, women formed groups and alliances where they had space to express their desires and dreams for social recognition. The majority of them were wealthy women who had established charitable foundations and started to publish journals for women. A major breakthrough occurred when in 1930, they acquired the right to vote under certain circumstances, such as to know how to read and write (Dalakoura and Ziogou-Karastergiou 2015). This was coupled with the growing numbers of women workers during the interwar years (Avdela and Psara 1985). Having the right to study and work made her an active and productive member of society and strengthened the self-esteem of the Greek women. As mentioned before, the role of the Greek woman began to change more effectively since 1952, when she acquired the right to be elected in parliamentary elections. Consequently, the role of the mother changed as well, since it had to be combined with that of a working woman. The modification of woman's role during the second half of the twentieth century was crucial to family life since she gained important legal rights and was emancipated. As was widely known, after entering the workforce, the traditional model of the housewife broke down and was replaced by a more multidimensional role. Women were absent from their house more hours during the day and had to let their children be raised by somebody else. Furthermore, the traditional Greek family model was reshaped—firstly, due to innovative medical advances, such as in vitro fertilisation and sperm banks, and secondly, due to societal changes, such as feminism and urbanisation (Barmpouti 2015a). As a result, the change to family law in the mid-1970s was a reasonable outcome of the existing situation. As elsewhere, the establishment of women's clubs and societies at the time also reflected the fact that feminism became stronger (Rendall 1985; Frevert 1989). However, gender equality was only acknowledged by the state almost thirty years later. In the meantime, women entered the workforce and academia; they elevated their social status and became actively involved with politics (Henig and Henig 2001; Inglehart and Norris 2003).

The history of post-war Greek eugenics illuminates the participation of female experts in its development. Not only was Maro Kanavarioti the representative of the HES abroad, but well-known professionals and academics were involved in its activities. For instance, Lina Tsaldaris was one of the founding members of the HES and the NUSE and represented Greece in the IPPF's meetings with Louros as Honorary Associates. She was the most representative example of a woman who made great efforts for child protection in Greece. She supported women's clubs and represented the country in international organisations, such as the UNICEF. Tsaldaris managed to become Minister for Social Care in the Greek Parliament and the first female minister in Greece. Another example was the gynaecologist Popi Spelioti-Bazina. Bazina was President of the Association of Female Greek Scientists and member of the HES. Judging from her writings, Bazina was a feminist who struggled for female emancipation and gender equality. Most importantly, Spelioti-Bazina was one of the few women who deliberately published articles on eugenics and birth control.

Considering that women had access to education and job opportunities, much of their time was spent on their activities outside the home. This automatically meant that her role as housewife was only part of the new multidimensional role. One of the first alterations in family life was the postponement of a woman's role as a mother. As Trichopoulos had pointed out, in the 1950s and 1960s, women tended to marry later than previously resulting in later childbearing. As a consequence, the reproductive years and the number of children diminished; then, large families gave way to smaller ones. Moreover, there was an observed willingness to control reproduction and have access to contraceptive techniques, either amateurish or professional. Gynaecologists, such as Louros and Panayiotou in Athens and Tsacona in Thessaloniki, justified the desire of their patients to learn how to plan their families and avoid unwanted pregnancies. The huge number of unwanted pregnancies and abortions was also the result of the total absence of sex education in schools or elsewhere. As will be portrayed in the analysis of the conference on sex education organised by the HES, not only was sex education in Greece non-existent, but also efforts towards its implementation were limited and often prohibited either by parents or by teachers. The lack of sex education was an important deficiency of the reproductive health and choices of Greek women.

In the late 1970s, the HES devoted three conferences to the institution of the family. The selected time period was not at all accidental, but echoed the social progress. The first and most thorough public discussion of the new family law was organised by the HES in 1976: 'The family today and tomorrow' (Hellenic Eugenics Society 1976b). Later two more conferences, one in 1978: 'Legal problems from the point of view of medical sciences' (Hellenic Eugenics Society 1978a) and another in 1979: 'Parental Authority or Care' followed, but only to examine the legal aspects of family law (Hellenic Eugenics Society 1979).

The discussion at the 1976 conference was interdisciplinary, including perspectives ranging from pedagogy to theology. The theological view could not be missing from a discussion on family in Greece. The Greeks were traditionally strongly attached to the Orthodox Church and its Christian morality. Savvas Agouridis, a Professor of Theology at the University of Athens, presented the Christian perspective on marriage and the conjugal relationship. Marriage was the first step towards the foundation of a family, which was translated into the 'completion' of a human being and the continuity of the human species. Agouridis explained that according to the Holy Bible and Christian tradition, the institution of marriage was regarded as a highly respected relationship between a man and a woman that cannot be spoiled. However, from the Christian eschatological point of view, no human relationship was final. The new life that the future Kingdom of God dictated would be beyond human relationships, as these were perceived and experienced by humans. This new state of being would not be humane, but a situation where man would acquire God-like characteristics and surpass his nature. Agouridis focused on the Orthodox perspective of marriage and family, as it was experienced in Greece. In the Greek tradition, marriage and family were of great importance. The foundation of a family was regarded as the main purpose of life. Unmarried people were considered incomplete and sometimes even marginal. This conservative thinking gradually changed, simultaneously with the change from the extended family to the nuclear. Nuclear families replaced the large families of the past and this was a matter of concern, because, according to Agouridis, the new family model was not as stable and cohesive as the old one. According to Agouridis, the traditional model offered psychological security and stability, whereas modern nuclear families which were built in stressful, city centres retained loose bonds among its members and could be disrupted easier.

The sociological view of the institution of family was successfully presented by Artemis Emmanuel. She began with the sociological principle that each society was a network which included many subsystems, one of which was the family. As in every relationship between a wide system and its subsystems, a mutual feedback was observed between the work of the society and that of the family. Consequently, the foundations and functions of society were often reflected in the family, and conversely, the activity of the family unit influenced the motion of society. In particular, from the beginning of the twentieth century, the Greek family had to confront a number of difficulties, some of which were national insecurity, immigration and emigration, financial inadequacy and a poor educational system. Emmanuel quoted Valaoras' observations on Greek demography, which had shown that this uneasy situation of the Greek society led to demographic stability and population ageing. The sudden urbanisation of the neo-Greek society in big urban centres, such as Athens, was followed by a passive imitation of foreign, Western ideals. As Emmanuel argued, the majority of the new Greeks, who inhabited the cities, pursued a fake cosmopolitism, which became threatening for the national and cultural identity of the country. Following the previous explanation of the relationship between the family and society, the imported lifestyle models were so influential to the family models that they contributed to the shift from familism to individualism. Therefore, the traditional Greek family model, along with the values and principles it represented, was finally thrown into disarray and the nuclear family model prevailed.

However, Emmanuel was optimistic about the future of the family. On the one hand, she argued that the foundation of a family was an innate characteristic of human beings; on the other hand that in every society there was a family model, which was transformed in accordance with the societal changes. Therefore, it was possible that a positive development in society would improve the family. Social progress and scientific development could improve the quality of life gradually, in both individual and collective levels. To this end, a social agenda based on science, technology and moral values was absolutely necessary. In this context, family planning was essential, because the role of the family was significant to society. Moreover, the return to the older, traditional role of the family was essential to fulfil this purpose. Emmanuel also mentioned that genetics, eugenics and sociology should be aware of the problems that arise from the new family models, namely the technologically engineered families such as those resulting from sperm banks or in vitro fertilisation. She acknowledged, though, that these scientific advances primarily assisted the institution of the family in fulfilling its psychosocial and cultural role. According to Emmanuel, Greek families, on the one hand, should keep their authenticity and their national identity, but, on the other hand, should be incorporated into the European idea of a unified but pluralistic society.

As is widely known, every social change should be under a legal protection in order to be safeguarded. The above-mentioned condition of the Greek society during the twentieth century had to be legally secured. In fact, as Skorini-Paparigopoulou, a Professor of Law at the University of Athens, explained, the legal system usually follows a social change, not the opposite. This time lapse was called 'cultural lag' and reflected the delay in legalising a social fact. In the case under examination, the legal response to social change was article four of the constitution about equality between the sexes. The advanced position of women in the family and society had already existed, but it needed to be legally acknowledged.

Later, during the conference 'Parental Authority or Care' (1979), Kalliopi Spinelli, a sociologist of Law, added that the modification of family law in accordance with Article 4 would not introduce anything new, but would adapt its outdated legal provisions to modern society. The legal framework was anachronistic; it did not follow the contemporary social reality of the institution of the family, which was formed in the technologically developing society of Greece. Maria Fatourou added that Greece should follow the example of other European countries that changed their family law in the past five years, because Greece belonged to Europe (David 1992; UN 2002; Barmpouti 2015a).

The conflict between modernity and tradition was mirrored in many aspects of the family life, including childrearing. From a paedagogical point of view, Evangelos Papanoutsos argued that the change from parental child rearing to child raising by grandparents or nannies was the most problematic. In the past, the children did not leave home before primary school; they learned the first elements of knowledge inside the familial environment. In the modern society, the children were raised by a person, all too often, from outside of the family circle, starting around the time of breastfeeding. Nannies and babysitters took the place of parents and the paedagogical role of the family failed. Therefore, according to Papanoutsos, the state had to enrich the educational system for its future citizens.

The, then modern, social conditions did not benefit the cohesion of the family. As a psychiatrist, Georgios Christodoulou presented the psychological side and a possible reaction of the children. The emotional bonds between children and their parents had become so loose that in many cases they faced serious psychological problems. Christodoulou explained that the lack of a good familial environment caused children to experience disturbances in behaviour, speech and personality. In extreme cases, there were studies that supported the idea that people raised in problematic families tended to have criminal behaviour. It was not a coincidence that many psychoses were attributed to the bad relationship that the patient had with his familial environment. Christodoulou quoted the theories of Sullivan (1968), Lidz and Lidz (1949), Singer and Wynne (1966) and the double-bind theory of Bateson (1972) to justify his position. It is worth mentioning that these theories formed the basis for the anti-psychiatry movement, which demonstrated the link between a problematic social environment and the development of psychoses.

3   EDUCATION

3.1   Sex Education

The originality of the HES's conferences was not limited to eugenics and population problems, but included the thorny issue of sex education. Although the HES discussed the subject in 1963, the time was not appropriate to produce significant outcomes. There were negative reactions and doubts about the effectiveness of the addition of sex education as a separate course in schools. The prevailing Greek perspective was that the family should play the role of the educator in sexual matters (Ioannidi-Kapolou 2004). As was previously demonstrated though, the educational role of the family was limited or non-existent. The absence of sex education was part of the wider problem of restricted knowledge of reproductive health and hygiene, which resulted in the large number of unwanted pregnancies, abortions, single mothers, the spread of venereal diseases and limited use of contraception. Up to the present day, sex education has not been part of the Greek schools' curriculum. It is surprising that in a country with a significant problem of numerous induced abortions, sex education has been ignored.

Although there were teachers and scholars who voiced the necessity of sex education in schools, no progress had been made until 1979 when the HES organised a two-day, interdisciplinary symposium on sex education, illustrating many aspects of the topic and referring to the obstacles that impeded its inclusion in the school curriculum (Antonopoulos 1953).

Alexandros Stavropoulos, a theologian, was responsible for the organisation of the symposium and delivered the keynote speech. The minutes of the conference were published by the HES in 1981 (Hellenic Society of Eugenics and Human Genetics 1981). The then President of the HES, Ioannis Danezis, claimed that the published volume aimed at filling the gap of sex education in Greek scholarship. He added that the country was prejudiced against sex education, which was true. Therefore, that volume would be a useful tool for those who supported the dissemination of sex education and worked towards its materialisation. Except from the minutes of the three sessions that comprised the symposium, the volume included a list of addresses and telephone numbers of centres, organisations, public services and journals for sex education related directly or indirectly to sex education in Greece, UK, France, Switzerland,

Belgium and Germany. Moreover, there was a thematic bibliography which included a list of dictionaries, encyclopedias, book series, journals, research studies and audiovisual material, handbooks, institutions, as well as information on international opinions for sex education. It also included information on special subjects, such as sexual anthropology, contraception, unmarried mothers, venereal diseases, marriage, women, abortion, family planning and sterility. This was a unique companion for sex education which provided the reader with unprecedented information about the subject in Greek and international contexts.

The papers were prepared in advance by three working groups, each dealing with a different aspect of sex education. The first one chaired by G. Maniatis discussed the human sexual life under the prism of sex anthropology and biological, psychological, sociological and theological approaches. The second one chaired by M. Kinigou dealt with the international presence of sex education in comparison with Greece. The last group, chaired by I. Markantonis, prepared the discussion for the possibility of the inclusion of sex education in the Greek schools' curricula.

Cleopatra Oikonomou-Mavrou, a Professor at the National School of Hygiene, portrayed the condition of sex education in Greece. She thus explained that it was neither prohibited nor encouraged by the state. Although there was no legal constraint for its implementation, prejudice impeded it. The absence of sex education led children to obtain indirect and often non-scientific information about sexual affairs, mostly from their peers, their parents or printed material. Oikonomou-Mavrou identified the reluctance of teachers to undertake the responsibility of sex education. Simultaneously, teachers lacked training in teaching such subjects as reproductive health, sex, contraception or family planning, because colleges and universities, even the medical school, did not include sex education in their curricula. However, the Orthodox Church was actively preoccupied with the subject and often organised relevant lectures and meetings about the preparation of adolescents for marriage, procreation and familial life.

From the secular perspective, the only example of premarital advice was the Premarital Advisory Centre at Alexandra Maternity Hospital under the direction of the President of the HES, Danezis, and the participation of Valaoras and Kanavarioti. The establishment of the Centre was initiated by the HES and partly funded by the Ministry of Social Affairs (Stavropoulos 1970). As mentioned before, its function lasted only for a couple of years (1966–1968), because it was an experimental institution aiming at evaluating the situation of premarital and conjugal relationships of the Greeks. The ultimate goal of the project was to take advantage of the results of the function of this Centre in order to establish an official premarital and conjugal advisory institution. Among the reasons for establishing such an institution was to confront the low birth rate and the incidence of unwanted pregnancies and abortions, and the prevention of divorces and venereal diseases. However, their target was not realised after the closure of the experimental centre. Similarly, sex education was not disseminated by any official institution. Much later, the initiative to publicly disseminate family planning advice was taken by a non-governmental organisation established in Athens by a group of volunteers under the leadership of the gynaecologist, Dr. Kintis, and a member of the parliament, Mrs. Tsouderou in 1976 (Ioannidi-Kapolou 2004).

Mass media and publications for sex education were also scarce; only a few books, which were translations of foreign ones, were published. Remarkably, Oikonomou-Mavrou claimed that paediatricians, who asked radio stations to include brief messages or interviews about sex education, experienced disapproval and rejection. Oikonomou-Mavrou also identified the widespread belief that sex education would encourage children and youngsters to begin their sexual life earlier than 'normal'. Fear of premature sexual activity caused the majority of Greek society, including parents, teachers and health professionals, to oppose sex education.

As already discussed, the role of the Orthodox Church was not pervasive, albeit decisive to the life of the Christian. The Orthodox rhetoric in favour of sex education was based on the belief that man is a psychosomatic union. Thus, the physical entity of man cannot be ignored by the Church. According to Fouskas, a priest of the Greek Orthodox Church, the Church should be actively involved in sex education because the Christian does not blindly obey the commands of the priest, but demands argumentative discussion and education. Similarly as with schools and universities, the appropriate education of the clergy in order to confront the problems of teaching and advising about sexual matters was central to Fouskas' argument. Profound study and appropriate methodology were deemed necessary for a successful education. In this context, Fouskas claimed that the Orthodox Church should publish one or more encyclicals such as the *Casti Connubii* of the Catholic Church and the *Problems of Marriage and Divorce* by the Archbishop of Canterbury (Casey 1960; Stokes 1972; Rosen 2004; Löscher 2007; Leon 2013).

Summarising the general outcomes of the symposium, it was unanimously argued that: sex education was necessary at every age, with an emphasis to childhood and youth; parents should cooperate with teachers in order to assist the child during its psychosexual development; and particular attention should be paid to the selection and training of sex educators, in schools, churches or other institutions. Above all, the implementation of sex education courses should result from a coordinated action by the family, educational, religious and state institutions. The HES offered the expertise of its members and bibliographical and audiovisual material at the disposal of every interested agent or institution and the state.

3.2   Education for Health Professionals

As was discussed, the Working Committee of the HES made efforts to organise lectures for both health professionals and non-academics. It was essential to reach both target groups because they aimed at spreading the eugenics knowledge across the society irrespective of social status or level of education. Thus, each subject was adjusted to the audience. Given that many eugenicists were physicians, in particular professors at the Medical Schools of Athens and Thessaloniki, one can easily assume that lectures on eugenics were unofficially took place during the regular schedule at the universities, too.

In 1963, the HES inaugurated a series of scientific conferences targeting primarily physicians. It was part of their plan to educate health professionals about issues of eugenics. The first solely medical round-table conference, organised by the HES, was on the subject of 'The harmful influence of various factors on embryogenesis'. The minutes of the conference were entirely published in the medical journal *Iatriki* by the Society of Medical Studies (1963). The participants, who were all physicians, discussed physical, pharmaceutical or chemical factors that could have negative outcomes in pregnancy. Papers included: 'The harmful influence of external and inherited internal factors on gene cell and the embryo' by Louros, 'The morphological elements of reproduction' by Papatheodorou, 'The elements of physiology of the reproductive system and harmful influences on the gene cells and the embryo' by Danezis, 'The hormonal negative effect on the embryo' by Batrinos, 'The importance of pharmaceuticals on the induction of defects to the formation of the embryo' by Moiras, 'The effect of the maternal infection to the embryo' by Papadatos, 'The influence of the ionic radiation on gene cells' by Pontifikas and 'The congenital diseases caused by radiation on the embryo' by Kostaridis.

The conversation which followed the end of the presentations was equally important because many important physicians expressed their views on the subject. Among them was Konstantinos Choremis, Eminent Professor of Paediatrics at the University of Athens and founding member of the HES who congratulated the President of the HES, Louros, for the initiative to organise such a conference on the harmful effects of chemical and other pharmaceuticals during embryogenesis. Choremis defined the remedial role of eugenics to the prevention or modification of external factors after conception, because the discovery and prevention of harmful environmental factors were more promising and plausible than the discovery of genetic factors. He claimed that congenital diseases were only partly confronted by the medical advances and prenatal tests. Choremis was very critical of pregnant women who took medication without any restraint. He remarked that 'patience and pain seem alien to human nature nowadays that people exploit scientific advances more than is necessary'. The role of eugenics should be to educate pregnant women and help them to avoid such irresponsible behaviour and pharmaceutical abuse. He propounded that: 'modern dysgenics and the multiplication of mental illnesses are more the result of modern civilisation; the work of human, and less the work of nature. Eugenics should aim at the prevention of harmful and dangerous effects on human behaviour'. In conclusion, Louros suggested that the research of environmental harmful effects to embryogenesis was very important and should continue to progress. However, in most cases, there was a genetic predisposition. Therefore, the manifestation of malformation was multifactorial. Louros also focused on the education of the gynaecologists in saying that it is their responsibility to inform and protect pregnant women.

Putting that conference in the medical historical context, it must be acknowledged that it was a pioneering work for the Greek medical community. On the one hand, during the 1960s, many new medicines were released on the market but on the other hand many physicians did not know how to prescribe them correctly. Furthermore, the issue of polypharmacy was tormenting Greek society and affected pregnant women that used to take unnecessary medication without prudence. As was commented by Louros, the conference lasted four hours and the audience was large. Aside from environmental influence, another matter of concern for the pregnant women was the transmission of hereditary diseases.

Although genetic determinism was popular among physicians and biologists, many health professionals urged the need for prevention from environmental, harmful factors during pregnancy that cause birth defects. Experts have admitted that the advantage of this kind of preventive medicine or eugenics during pregnancy was the ability to avoid, control or eliminate the presence of external damaging factors, such as radiation, consumption of chemical drugs and maternal infection, which could lead to malformation or injury.

4   Hereditary Diseases

4.1   Premarital Medical Examination

Blood examination before marriage was a topic of discussion in the Pan-Hellenic Medical Conference organised by the Medico-Chirurgical Society in 1958. During the conference, the prevention of hereditary diseases to secure good progeny was the prevailing opinion. Professors of Cardiology and Pathology emphasised the disastrous repercussions for family, society and race resulting from the marriage of unhealthy individuals. Katsilamprou, a Professor of Cardiology, argued that the neurological examination should be added to the laboratory examination in order to avoid the birth of epileptic children. Even cancer predisposition was attributed to a mother's deficient heredity, being transmitted through breastfeeding. Thus, the premarital health certificate was deemed absolutely necessary by Katsilamprou (E. M. 1958). The lack of knowledge of genetics and the mechanism of disease transmission was obvious in the given example. Before the systematic study of human genetics, the eugenics rhetoric was based on hypothesis. However, the eugenic goal for healthy reproduction at any cost for the benefit of the family, and in the long run for the society, was prevailing among physicians without considering the absence of solid scientific arguments. Similarly to the beginning of the eugenics movement in the beginning of the twentieth century, eugenics arguments were based on sociopolitical beliefs rather than scientific results. In the same context, the premarital medical examination for the prevention from the birth of unhealthy descendants was a matter of eugenics since its beginning. Diagnosis was secured by blood examination as the only available means before the prenatal genetic test was added. The premarital medical examination would not produce any important medical results but was often obligatory.

In Greece, although the discussion about premarital health certificate was vivid since the beginning of the twentieth century (Trubeta 2013), it became compulsory only during the years of the dictatorship (1967–1974) with Law 300/1968 (Official Government Gazette 1968). After being legally imposed to the prospective spouses, every couple was obliged to provide it to the authorities in order to get married. The results of the medical examination, however, were confidential and the physician was protected by the law. In reality, the certificate of examination had not the form of a declaration of good health, but it was verification that a medical examination took place. This suggests there were no official means for state intervention in marriages and procreation, with eugenic marriage guidance mostly occurring in private practice.

This certificate was voluntarily given to couples who wanted to be examined before marriage. They visited a doctor to whom they provided the necessary information about their family's medical history and they were also examined themselves. This examination occurred in two parts. One part was the actual examination and the other was the examiner's advice in case of an undesirable result.

Apart from including the premarital health certificate in various discussions, the HES discussed it in detail during its conference: 'Premarital medical examination' (1978b), in order to evaluate its usefulness ten years after its legal implementation. The members of the symposium were unanimously positive towards the voluntary character of the examination, but negative towards the compulsory one. Their arguments were more based on preventive medicine in this context. In fact, Danezis admitted that globally the premarital medical certificate was not compulsory. Furthermore, many states proposed the establishment of special genetic centres, where the examination and advice would be absolutely voluntary, such as the experimental Premarital Advisory Centre at Alexandra Maternity Hospital (1968) and the Centre for the Prevention of Mediterranean anaemia in Athens (1974) (Hellenic Eugenics Society 1978b).

Chaniotis, Director of the Ministry of Social Services, analysed Law 300/1968 and explained its features. Firstly, the premarital medical examination became obligatory for those who wanted to get married legally. Secondly, the certificate could be obtained by the couple only after examination. Thirdly, in the event of an unwanted result, this would not be written on it. The purpose of the certificate was exclusively to show that the examination took place. This was the reason why it was named 'Certificate of Medical Examination' (*Πιστοποιητικό Ιατρικής Εξετάσεως*) and not, for example, 'Health Certificate' (*Πιστοποιητικό Υγείας*). A fourth point was that the physician should be absolutely discreet. There were penalties, if they transgressed the medical confidentiality. Furthermore, in the case of a diseased person, the physician was obliged to inform the patient about all the details of the disease; but let the patient take the final decision. The patient alone was the person responsible for the decision of whether to get married and have children or not. The decision should be made independently. Last but not least, the examination was free of charge when the couple was examined at a public health institution or at their cost if they wanted to visit a private physician. In general, the diseases that were considered dangerous were mainly infectious diseases and not just congenital, such as leprosy, tuberculosis, syphilis and psychological disorders. The law permitted additions and exclusions in this list (Hellenic Eugenics Society 1978b). A problem that came up was the possibility of one spouse hiding the disease from the other. As was shown, someone could obtain the certificate claiming that he is healthy, while he was diagnosed with a disease. The presence of the document could provide false evidence of the person's health. Therefore, the premarital health certificate was deemed ineffective and misleading.

From the medical point of view and as an expert in Mediterranean anaemia, Fessas underlined the fact that it could prove dangerous, because someone could choose to get married and have children despite the fact that he was diagnosed with a congenital disease. Therefore, this couple could give birth to defective children intentionally. As for the safety of the diagnosis, Fessas claimed that there were a large number of diseases that could not be accurately diagnosed, whereas there were others like sickle-cell anaemia and Mediterranean anaemia that could be diagnosed safely.

From the legal point of view, Kassimatis, a Prominent Professor of Constitutional Law, explained the potential harmful repercussions of Law 300/68. He used the 'slippery slope' argument to question the limits of state intervention on the individual for reasons of positive and/or negative eugenics. Moreover, he wondered about the presuppositions that the lawmaker based his guidelines of the eugenic medical examination. He referred to the vagueness of the third article of Law 300/68, which gave the right to the state to impose some prohibitions in case of undesirable medical results. These were the prohibition of marriage for a certain time period or forever, which he thought was an insult to human dignity. According to Kassimatis, the atrocities of National Socialism in Germany were made because this political party wanted to impose their politics via hygiene programs and laws—not to protect society from bad progeny. In order to prevent society from the repetition of the above example, he proposed that two fundamental principles that were stated in the Declaration of Human Rights be respected. The first one was the respect of human dignity; the state should not intervene in people's personality. The second was the principle of free expression; each person had the right to use social institutions as they wished—in this case, the institution of marriage. Based on these principles, every examination which aimed at negative eugenics, such as the prohibition of marriage, should be banned as unconstitutional. Kassimatis claimed that while a system of eugenics should be adopted by the state, in order to prevent the spread of the congenital diseases, this should be structured upon the respect of human liberties. Preventive medicine was practiced for the prevention from the spread of hereditary, venereal or infectious diseases by providing recommendations, not by imposing compulsory examinations. Therefore, the prohibition of marriage and procreation was unethical and unconstitutional, but genetic counselling was recommended.

To this end, Kattamis, a physician and another Mediterranean anaemia expert, proposed a system of pre-marriage counselling aiming at the creation of healthy families from the physical, spiritual and psychological view. Apart from the prevention from congenital diseases, the premarital advice should point at the information for the dangers of the embryo and its protection. In some cases, the physician should extend his contribution to matters of fertility, procreation and family planning. The first stage of advice should be information, the second safe laboratory examination and the third and most important should be the proper guidance of the couple.

In fact, the word 'guidance' was not accurate, because the physician in such cases had to be as neutral as possible. The role of the advisor was to analyse and explain the health condition of the examined individual in order to help him to reach a decision about whether to get married and have children or not. The physician should hide nothing from the patient and try to be very informative in order to enlighten him (Toth et al. 2008). Fessas, as a physician, admitted that it was very difficult to be absolutely neutral because most of the time the patient asks for a physician's advice and because of the invasive nature of the profession.

It would be easy for a physician to impose his opinion as the right one, but when acting as a genetic counsellor, he should only be informative and neutral despite his ability to influence the patient.

As far as psychological disorders were concerned, there was a conflict between Christodoulou, a psychiatrist, and Kattamis, a physician. On the one hand, Christodoulou complained about the lack of information regarding the advice to be given to psychotic patients. Furthermore, he discussed the case of schizophrenia and argued that everyone had a possibility of up to 1% to develop this disease; if one of their parents was schizophrenic, then the percentage would be 11%; if both parents were schizophrenic, then they would have a 45% possibility to develop this disease. As a result, not only was it important to know the way a disease was transmitted, but also the damaging experience of a child who lives in a psychotic environment. Maybe schizophrenia was not transmitted genetically, but it should be examined as well. Kattamis, on the other hand, insisted on the fact that psychosis could not be proved genetically, in a laboratory. It could develop after thirty years or more. Therefore, it could be reckless to adopt certain rules of advice for those cases. Fessas added that the psychiatrist, not the physician who would perform the premarital examination, should advice a psychotic patient.

To sum up, there were some common conclusions that all agreed with. First of all, every prospective parent should be responsible for their actions regarding reproduction and should visit a physician who could help them do so. Therefore, they insisted on the importance of medical counsellors, who were supposed to explain in detail the medical problem and give useful medical advice to the couple. The new couples should be aware of the dangers that threatened them and their offspring. Furthermore, the members of the HES emphasised the difference between the preventive character that such an examination entailed and the constant eugenic control of the nation by the state.

Family planning and the premarital health certificate were eventually re-defined by Law 1036/80 in 1980. This permitted the foundation of Family Planning Centres and simultaneously abolished the compulsory premarital health certificate. Remarkably, Law 1036/80 was signed by the then Minister of Social Services, Spyros Doxiadis, former President of the HES (Official Government Gazette 1980). The Hellenic Society of Family Planning, Contraception and Reproductive Health was active in Greece only after 1976, and in 1985, it became an official member of the IPPF.[6]

4.2   Genetic Diagnosis and Selective Abortion

The HES paid particular attention to the diagnosis of a genetic disorder. In December 1975, it organised a round-table discussion under the title 'Antenatal Diagnosis' (Hellenic Eugenics Society 1976a). The approach was holistic and interdisciplinary, and the overall aim was to bring together academics of different backgrounds to exchange opinions and ideas regarding the issues posed by a genetic diagnosis. Based on the commentaries expressed by the participants, the following topics deserve attention: medical counselling, preventive measures, genetic policies, the option of abortion in case of genetic abnormality and the role of religion.

In his paper, Loukopoulos underlined the importance of proper medical guidance after a genetic test and diagnosis. The need for such guidance was necessary, he argued, mainly in three cases: when one of the parents had a congenital disease, when a child with a genetic abnormality was already born in a family where the parents were healthy and when the parents have undertaken a medical test which indicated high possibility of giving birth to a child with a genetic disease. The role of the medical advisor was crucial in this respect, although he/she had to base the diagnosis on two premises: confidence about the accurate diagnosis of the genetic disease of the parent or the child and its gravity as well as available information about the way of transmission.

Taking this argument further, Fessas added three more cases where medical intervention was necessary: when the disease was very frequent, when it was severe and when it was neither frequent nor severe, but lasted for a long period, thus also becoming a serious social problem. One such genetic condition was considered to be Down's syndrome. Fessas then highlighted that there was still insufficient knowledge about the so-called bad gene, except those genes that caused serious illnesses. What was a defective genetic disposition the one day could be perfect the other, he argued. Often, scientists were not able to offer a definite answer and a safe choice to the public.

Fessas considered the role of the physician and the impact of the diagnosis on the patient equally important. He argued that scientific advances influenced the function of society. People should be aware of the new technologies in medicine along with their use. Fessas claimed that people should not be tempted to alter their genetic inheritance for eugenic reasons and that scientists ought to allow biological variety in society.

The psychiatrist, Konstantinos Panagiotakopoulos, described the psychological problems caused by a negative genetic diagnosis. Many people who confronted such problems needed the help of a specialist and proper medical guidance. A negative diagnosis not only affected the parents, but also the wider family circle. Panagiotakopoulos then discussed the role of the genetic advisor who could be the family doctor. Being in this position, the family doctor had to be compassionate but remain neutral and try not to influence the parents when making a decision. The physician should only help the parent decide and not impose his own beliefs. In many cases, though, this was not possible, Panagiotakopoulos conceded. Genetic diseases not only affected the individual and his family but caused social problems as well. With this consideration in mind, genetic advisors often prompted the parents to make the, presumably, correct decision.

In agreement with Panagiotakopoulos, Eleni Marouli, a social worker, argued that genetic counselling should be neutral but very informative, so as to be helpful to the couple. It was the physician's responsibility to bring about equilibrium between the couple and to ensure that there would be a good relationship between the couple and the rest of the family circle. Genetic counselling, she suggested, should consider every patient individually. Each person was different and unique; therefore, the genetic counsellor should be flexible and caring. Marouli added that the genetic defect was perceived in various ways according to its external manifestation; the level that affects the patient's social life; and society's behaviour towards the affected individual. In this context, Marouli pointed out the psychological repercussions of a negative genetic diagnosis for the life of the couple. When someone knew that he or she was a carrier of a genetic disorder, they frequently became insecure, frustrated and generally shaken. The reaction to such a diagnosis varied according to the individual's cultural and educational level, religious beliefs, etc.

Danezis was the only participant who mentioned the other side of genetic testing: positive diagnosis. If the test was positive, the parents were generally not anxious about the health of their child, particularly when they already had an 'imperfect' child or when there were recorded congenital diseases in their families. Danezis thus described genetic testing as a method of prevention and as a means of stress relief for prospective parents.

Genetic testing as a method of prevention was also raised by other participants. According to Dimaki, health improvement could be accomplished in three ways: firstly, early diagnosis of a genetic abnormality; secondly, the prevention of conception of defective children; and thirdly, selective abortion. Dimaki emphasised the second option, in particular. She believed that prevention was better than cure, so everybody should focus on the methods of prevention. The methods she suggested were the following: selection of spouses on a rational basis according to their medical record, the permanent use of contraceptives or even voluntary sterilisation in case of negative diagnosis of one or both spouses, or in vitro fertilisation using a healthy donor. She admitted that the above recommendations were going to elicit negative social reactions, which depended on the social structure and the dominant social values of each society. Dimaki argued that the disciplines of sociology and biology should meet at some point, because their cooperation would provide solutions to the problems of eugenics. Both disciplines should find a way to secure the socio-biological betterment of mankind. She claimed that both sciences interact with each other and have a common goal, which is eugenics. On the contrary, Marouli focused on education and suggested sexual education and courses of family planning in schools, educational television programs and the continuous education of the specialists, such as physicians, social scientists and educators, as effective, preventive measures.

The most important factors to take into account in order to tackle the hereditary diseases, Kattamis argued, were the disease's frequency and gravity, lack of therapy, effectiveness and the cost of preventive measures of each disease. Therefore, population studies of the congenital diseases in Greece were imperative. Kattamis focused on three congenital diseases; Down's syndrome, which was also associated with the age of the mother; Mediterranean anaemia; and sickle-cell anaemia, which frequently appeared in Greece. Constantinos Crimbas for instance would accept eugenic policies for Mediterranean anaemia, sickle-cell anaemia and maybe a medical intervention for Down's syndrome. He made clear, though, that the decision should be personal and not after state intervention; the state should only provide the person with the relevant services. In addition, Kattamis was absolute about the urgent necessity of abortion in the case of such a diagnosis, because often the only available means of tackling a disease was selective abortion.

Moreover, Crimbas claimed that medical and biological advances offered the opportunity to establish and apply measures of genetic policy either individually or governmentally. He explained that these policies could be divided into positive and negative eugenic policies. The negative were translated into the effort to avoid the presence of pathological phenotypes; the positive were the effort to multiply the positive hereditary traits, based on a system of selection, like animal breeding. He claimed that only some of the negative eugenic policies should be adopted by the state, not positive ones, in the fear of a repetition of the Third Reich's atrocities. Crimbas suggested certain measures, in the event that both prospective parents were carriers of a hereditary disease: (a) to prohibit their marriage, (b) to let them get married and reproduce, but to examine the embryo and propose selective abortion if it was defective, (c) to let them get married, but either to decide by themselves or to be prohibited by the state to have children, and (d) to let them get married, but to have only the choice of in vitro fertilisation using a donor.

From the biological point of view, Crimbas admitted that these measures would not lead to genetic purification, because the diagnosis was not always accurate and the knowledge regarding the transmission of disease was not always clear. Only a slight biological change could appear by adopting these policies. Crimbas finally suggested that the optimal solution was to allow the couple to get married but to abstain from procreation. He argued that eugenic policies could lead to the breeding of people, who do not suffer from a congenital disease, but it could not alter the genetic pool of a population; this could not be genetically enhanced.

Regarding selective abortion, Simopoulos focused on the possibility of the birth of a defective child. He admitted that once the prospective parents were informed about the health condition of their child, they were responsible for the continuation of pregnancy. The psychological and financial burden of this decision was heavy and important concerning both their own and their child's future life.

Although today there are medical methods of dealing with some genetic problems, at the time the most suggested solution was selective abortion. As Danezis noted, at that time there were only three methods of diagnosis during pregnancy: amniocentesis, intrauterine overview and placentacentesis, the last two of which were in an experimental stage. For example, Down's syndrome is not as severe a condition as it was in the 1970s. A large part of people with Down's syndrome have the chance of getting an education and living a 'normal' life.[7] However, there is still genetic and social discrimination against people with

Down's syndrome. Genetic counselling was often against continuing a pregnancy when the embryo was diagnosed with Down's syndrome. Lack of accurate information about a child with Down's syndrome, such as their average lifespan, often leads a pregnant woman to decide to terminate the pregnancy. Although it does not represent a eugenic policy, misleading information by medical professionals to influence a pregnant woman's decision-making might be a form of eugenics (McCabe and McCabe 2011).

Many eugenicists, including members of the HES, called selective abortion 'therapeutic' because it was recommended as a method of therapy in cases of genetic abnormalities. Danezis argued for the necessity of therapeutic abortion, although he recognised the lack of accurate diagnosis. He believed that genetic diagnosis of an abnormality must lead to the decision of therapeutic abortion. He explained that the above diagnostic methods could take place between the fourteenth and the seventeenth week of pregnancy, because then it was safe to interrupt a pregnancy in the event of a negative diagnosis. Although the results of the test would be more accurate if the test was taken later than the seventeenth week, the interruption of the pregnancy could be dangerous.

On the other hand, Stamatis distinguished, in legal terms, the life before and after birth. Human life—after birth—and health had intrinsically great value which made them the greatest natural and legal rights. The protection of life after birth was absolute and unconditional, whereas before birth it was relative. The legal approach was thus put on a different basis. Apart from the protection of the unborn child, there were legal problems that arose from a prenatal diagnosis, such as the responsibility of the physician who performed amniocentesis or abortion for reasons of eugenics. If amniocentesis caused the death or malformation of the child, it could not be regarded as murder, because it was done without the intention to kill or harm. However, before the examination took place, the physician should have excluded the possibility that his action could cause injury or death of the foetus.

At that time (1976), abortion for reasons of eugenics was prohibited. According to the Greek Penal Code, Law 304, paragraphs four and five, abortion was legally accepted only for the following reasons: the danger of life or health of the mother or in case of seduction, rape or incest. As a result, the physician could not legally suggest the interruption of pregnancy in any other case.[8]

The Christian Orthodox point of view was discussed by Alexandros Stavropoulos who repeated that genetic counselling should be inform- ative, yet neutral. The medical advisor should not make the decision on behalf of the couple. Nobody should decide on behalf of somebody else in spiritual matters, such as matters of life or death. Stavropoulos based his interpretation of genetic diagnosis on Christian anthropology. There were some fundamental values of Christian tradition that were outlined, such as the belief that man was created by God in his image, that man and woman were responsible for the transmission of life, and that procreation was blessed by God. According to Stavropoulos, the Orthodox tradition associated the sinful life with disease and bad prog- eny. Furthermore, in the ceremony of marriage was included the wish for good, beautiful and healthy children (υπέρ καλλιτεκνίας). Moreover, the Church cared about the good progeny and showed it practically by prohibiting the marriage between relatives and the prohibition of sexual relationships when a woman was menstruating, because it was believed that conception during menstruation would lead to the birth of children with genetic defects. The ideal case for the Orthodox Church would be if the conception was the result of a physical sexual relationship of the married couple, without the intention to avoid procreation, either by contraception or by interruption of the pregnancy. As already men- tioned, the only acceptable means of avoidance of procreation should be the abstinence of the couple. As for abortion, the Christian tradition was clear, and abortion was contrary to the Christian perception of life; it was considered as murder and an attempt against human life. On the other hand, the Church understood the difficulty of raising a defective child and had to be sympathetic towards those people who made the decision to interrupt the pregnancy for reasons of eugenics, when they came to Church with repentance. Dimaki, on the other hand, mentioned the latent eugenics expressed in the Christian prohibition of incest in order to retain the familial relationship out of sexual conflicts and rival- ries. She believed that behind it the effort to avoid the birth of defective children was hidden. Stavropoulos replied that the Church always sup- ported medicine and its curative role. Even if the respect of human life was above all virtues, the Church would not promote or allow the con- ception, which was predicted to give birth to defective children.

Dimaki, in agreement with previous presenters, repeated that the choices after a diagnosed defective genetic predisposition in one or both members of the couple were limited to the following: to avoid marriage;

to get married but avoid procreation with the use of contraceptives; in vitro fertilisation with a healthy donor; or to risk a pregnancy, but choose abortion for reasons of eugenics, in the event of negative prenatal diagnosis. Unlike most of the members of the HES, Dimaki went as far as to support the sterilisation of such a couple in favour of the rest of the society.

From the financial point of view, Petros Gemptos claimed that public expenses for health were a form of investment, due to the fact that they eventually offered prosperity and increased the value of human capital. Regarding genetic policies, Gemptos argued that when the cost of prevention from a congenital disease was lower than its future therapy, then, from a financial point of view, these preventive measures were desirable. He seemed to agree with Crimbas, who was cautious about new biomedical technologies and genetic policies. Gemptos believed that additional research should be done on influential factors of the health conditions. He thus said: 'Even if in the future we have the ability to test the impact of a defective gene, genetic policies should be applied only in states of emergency'.

Reminiscent of the aims of the HES set out in the preliminary meetings, Dimaki claimed that the state should be more active in matters of procreation. She suggested incorporating into its system the pursuit of the birth of healthy children. At that point, she mentioned the valuable contribution of sociology in a eugenic policy. Sociology could predict the social consequences of the application of preventive measures from the birth of unhealthy children; it could examine the reaction of different social groups in scientific advances, which related to family planning and finally provide the state with useful data regarding the ways of progressing public health without provoking social tension. Stamatis expressed the legal point of view and underlined that state eugenic policies should be limited by the constitutional freedoms of the citizens, because if we exceeded these limits, then a totalitarian ideology might appear.

4.3   The Case of Mediterranean Anaemia

The prevention of hereditary diseases was an essential component of eugenics. In Greece, as in the majority of Mediterranean countries, there was a growing concern for a particular disease, that of Beta-Thalassaemia or Mediterranean anaemia. Its name was attributed to the high percentage of carriers in the region, even though it was also detected in people of African and Asian descent among others (Cao and Galanello 2010). As was often expressed during the period from 1950s to 1980s, Mediterranean anaemia was the primal social and medical problem in Greece, justifying the special attention that was given to this disease. The HES discussed Mediterranean anaemia specifically during three of their conferences: blood and heredity in 1970, round-table discussion: antenatal diagnosis in 1975 and premarital medical examination in 1978.

Mediterranean anaemia is a congenital blood disease, which provokes blood disorders that escalate to a form of anaemia (Weatherall and Clegg 2000; Weatherall 2010). In its most severe condition 'homozygous Beta-Thalassaemia' (or thalassaemia major), the clinical symptoms varied from extreme anaemia to severe osteoporosis with spontaneous fractures, bone deformities and abdominal swelling (Kattamis et al. 1970). In the most common cases, the patients would require blood transfusions for the rest of their life. According to Fessas, a pioneer in studies on Mediterranean anaemia in Greece, medicine should keep these people alive because blood transfusions were the only thing that a patient should do. No matter how difficult such a situation might be; it was effective because people with Mediterranean anaemia had no other mental or physical problem, apart from a small number of red blood cells. Although he supported the above view, he argued that physicians should be obliged to recommend or impose preventive measures, such as a simple blood test to prospective parents. The matter of safety, regarding the accurate prognosis and diagnosis of the disease, was quite clear here. Although for a number of hereditary diseases the prognosis was not accurate for Mediterranean anaemia, it was safe.

Mediterranean anaemia belongs to the category of genetic diseases, which are not apparent in the prospective parents before taking the blood test, because the trait carriers do not manifest the disease. The only advantage of Mediterranean anaemia is that it can be accurately predicted even before marriage and conception. After taking the blood test, the prospective parents were able to decide about their future and bear the responsibility for taking the risk of giving birth to a genetically defective child (Modell et al. 1980). The percentage of transmission of the disease to the children of the carriers is the same as any other congenital disease, where the Mendelian laws of heredity were applied.[9] Moreover, a defective gene was expressed only when the person had inherited it from both parents.

There were numerous studies dealing with the incidence of this disease in Greece (Malamos et al. 1962). In order to better understand its range, Kattamis presented the results of studies between 1962 and 1972, which showed that in particular areas of the country, such as Euboea and the island of Rhodes Mediterranean anaemia reached 20% and sickle-cell anaemia reached 23% in areas such as Chalkidiki and Orchomenos. In 1974, Kattamis conducted research in the First Pediatric Clinic of the University of Athens regarding the number of children suffering from congenital diseases who were hospitalised, the number of days of hospitalisation and the number of the beds that they used. The results showed that approximately 21% suffered from Mediterranean anaemia. The percentage was extremely high and revealed the gravity of the problem. The second more frequent disease was sickle-cell anaemia and the third was cystic fibrosis (Hellenic Eugenics Society 1976a). Kattamis was convinced that the medical advances could be better appreciated with the cooperation of other sciences and the sympathy of the entire population in order to tackle the disease.

It is worth mentioning that Stamatoyannopoulos, Fessas, Kattamis and Loukopoulos were the founders of the first Centre for the Prevention of Mediterranean anaemia in 1975 in Athens.[10] The Greek state financially supported the function of the Centre and the campaign for the prevention of the disease. In particular, Loukopoulos claimed that when a problem took national dimensions, such as Mediterranean anaemia in the region of the Mediterranean Sea or the sickle-cell anaemia in people of African origin, then a genetic policy was called for (Loukopoulos 2011; Loukopoulos et al. 1983).

Genetic counselling dominated the discussions on Mediterranean anaemia (Barmpouti 2017). Fessas expressed the opinion that the most important medical recommendation was prevention by examination. Namely, all the couples about to get married should be examined. He claimed that the only possible solution for a couple who are both carriers of the disease was not to have children, because there was a high percentage of having a defective child. He suggested in vitro fertilisation with a donor or adoption, as alternative solutions. Timos Valaes, Director of the Institute of Child Health, agreed with Fessas in the prohibition of marriage when both parents were carriers but acknowledged that the measure was very strict since these people would still have 75% of giving birth to a normal child. He meant that they will have 25% possibilities to give birth to a healthy child and an additional 50% to have a healthy carrier, who had only one defected gene, so the child would be apparently healthy. As for the introduction of the examination for Mediterranean anaemia to the premarital certificate, Fessas argued that it would not be possible technically; each couple should take its own responsibility towards this problem. There were so many marriages that it was not possible to know if every couple was properly examined.

At the time, eugenics in Greece was intrinsically linked to family and procreation. The timeline of eugenics arguments begins with proper spouse choice, in terms of health and heredity, continues to eugenics during pregnancy and ends with proper childcare. Following this rationale, a variety of opinions of eminent Greek scholars and scientists for the legal protection of the child were examined: premarital medical examination, environmental influence during pregnancy and congenital diseases—with a particular focus on Mediterranean anaemia—selective abortion and genetic counselling. Despite the divergence of opinions and the variety of topics, one can claim that a consensus was reached on the value and effectiveness of preventive medicine.

Notes

1. See more about Banu's work in Marius Turda. 2009. "To End the Degeneration of a Nation": Debates on Eugenic Sterilisation in Inter-war Romania. *Medical History*, 53, 1 (January): 77–104.
2. Hellenic Eugenics Society. 1958. Minutes of the Executive Board's Meeting, Alexandra Hospital. *Louros Papers and Archive.*
3. Hellenic Eugenics Society. 1959. Newsletter of the Hellenic Eugenics Society. *Nikolaos Louros Papers and Archive.*
4. See Hellenic Statistical Authority. Greek population censuses: 1951, 1961, 1971, 1981.
5. Andreas Kokkevis was Minister of Social Services for the time period from 24 July to 9 October 1974. He thus participated in the Government of National Unity under Konstantinos Karamanlis, when also Nikolaos Louros became Minister of National Education and Religion. Spyros Doxiadis succeeded Kokkevis until on 21 November, the same year. See The General Secretariat of the Government. www.ggk-gov.gr/?p=1271. Accessed 23 May 2012.
6. See more details in: Hellenic Society for Family Planning, Contraception and Reproductive Health. http://www.oikogeneiakos.gr/index.php?option=com_content&view=article&id=79&Itemid=311&lang=el. Accessed 2 February 2014.

7. See Pablo Pineda, a Spanish actor with a diploma in Teaching and BA in Educational Psychology, with Down's syndrome in Aradas, Anahí. 2012. Los desafíos del primer licenciado europeo con síndrome de Down. *BBC Mundo.* http://www.bbc.com/mundo/noticias/2012/08/120823_cultura_pablo_pineda_aa.shtml. Accessed 23 March 2014.

8. Law 304 was replaced in 1986 by the Law 1906, which permitted the interruption of pregnancy in case of prenatal diagnosis showing severe abnormality of the embryo, which was going to result in the birth of a pathological newborn.

9. Gregor Mendel's laws of inheritance are the following: (a) the Law of Segregation; (b) the Law of Independent Assortment; and (c) the Law of Dominance. These are described in every book on biology and explain the transmission of a genetic trait. For a critique on Mendelian eugenics and anti-Mendelism, see Hamish G. Spencer and Paul, Diane B. 1998. The Failure of a Scientific Critique: David Heron, Karl Pearson and Mendelian eugenics. *The British Journal for the History of Science* 31, 4: 441–452.

10. Centre of Mediterranean Anaemia. Laiko Hospital Athens. http://www.laiko.gr/index.php?option=com_content&view=article&id=81&Itemid=125. Accessed 12 January 2016.

REFERENCES

Anon. 1958. A Pill Against Malthus' Prophecy. *Ikones* 147: 30–33.

Antonopoulos, Anastasios. 1953. *Sex Education of the Youth.* Patras.

Avdela, E., and A. Psara. 1985. *Feminism in Interwar Greece: An Anthology.* Athens: Gnosi.

Banu, G. 1939. *L' hygiène de la race.* Paris: Masson et Cie.

Barmpouti, Alexandra. 2015a. Population, Urbanisation and Eugenics in Athens, 1950s–1970s. *Revista de Anthropologie Urbana* 5: 73–81.

Barmpouti, Alexandra. 2015b. Eugenics and Induced Abortions in Post-War Greece. *Acta Historiae Medicinae, Stomatologiae, Pharmacie, Veterinae* 34 (1): 38–50.

Barmpouti, Alexandra. 2017. Genetic Counseling for Mediterranean Anemia in Post-War Greece (1950–1980). In *History of Human Genetics: Important Discoveries and Global Perspectives,* ed. Heike Petermann, Peter Harper, and Susanne Doetz, 462–483. Heidelberg: Springer.

Bashford, Alison. 2014. *Global Population. History, Geopolitics and Life on Earth.* New York: Columbia University Press.

Bateson, Gregory. 1972. *Double Bind: Steps to Ecology of the Mind.* Chicago: University of Chicago Press.

Blacker, J.G.C. 1954. Unchecked Populations: Some Comparisons of Rapid Growth. *The Eugenics Review* 47 (4): 235–244.

Burnova, E., and M. Garden. 2014. Naître à Athènes dans la première moitié du xxe siècle. Démographie et institutions. *Annales de Démographie Historique* 1 (127): 209–234.

Cao, A., and R. Galanello. 2010. Beta-Thalassemia. *Genetics in Medicine* 12 (2): 61–76.

Casey, T.J. 1960. Catholics and Family Planning. *The American Catholic Sociological Review* 21 (2): 125–135.

Connelly, Matthew. 2008. *Fatal Misconception: The Struggle to Control World Population*. Cambridge, MA: Harvard University Press.

Dalakoura, K., and S. Ziogou-Karastergiou. 2015. *Women's Education. Women in Education: Social, Ideological, Educational Transformations and the Female Intervention (18th–20th Centuries)*. Athens: Hellenic Academic Libraries.

David, H.P. 1992. Abortion in Europe, 1920–91: A Public Health Perspective. *Studies in Family Planning* 23 (1): 1–22.

Drakoulidis, Nikolaos. 1963. Consequences de la surpopulation sur la sante mentale et morale. *Acta Psychotherapeutica* 11: 464–472.

E. M. 1958. Blood Examination Before Marriage Is Necessary. *Ethnos*, September 15.

Fanaras, V.G. 2002. The Christian Orthodox and Other Churches and Religions View on Reproductive Issues. In *Proceedings of the Conference 'Marriage and Reproduction'*. Athens. http://www.ecclesia.gr/greek/holysynod/commitees/family/parousiasi_syzigia.html. Accessed 22 July 2017.

Finkle, Jason L., and Barbara B. Crane. 1975. The Politics of Bucharest: Population, Development, and the New International Economic Order. *Population and Development Review* 1 (1): 87–114.

Frevert, Ute. 1989. *Women in German History: From Bourgeois Emancipation to Sexual Liberation*. Oxford and Washington: Berg.

Galton, Francis. 1889. *Natural Inheritance*. London: Macmillan.

Goodhart, C.B. 1955. Natural Regulation of Numbers in Human Populations. *The Eugenics Review* 47 (3): 173–178.

Hellenic Eugenics Society. 1965. *Public Discussions*. Athens: Parisianos.

Hellenic Eugenics Society. 1974. Problems of the Elderly. *Iatriki* 11: 431–451.

Hellenic Eugenics Society. 1975. The Reproduction Problems of the Greek Population. *Elliniki Iatriki* 44 (3): 187–216.

Hellenic Eugenics Society. 1976a. Round Table Discussion: Antenatal Diagnosis. *Iatriki* 30 (2): 123–187.

Hellenic Eugenics Society. 1976b. The Family Today and Tomorrow. *Elliniki Iatriki* 45: 203–229.

Hellenic Eugenics Society. 1977. *Public Discussions*. Athens: Parisianos.

Hellenic Eugenics Society. 1978a. *Legal Problems from the Point of View of Medical Sciences.* Athens: n.p.

Hellenic Eugenics Society. 1978b. Premarital Medical Examination. *Materia Medica Greca* 6 (4): 299–315.

Hellenic Eugenics Society. 1978c. *Public Discussions.* Athens: Parisianos.

Hellenic Eugenics Society. 1979. Parental Authority or Care. *Nomiko Vima* 28 (2–3): 395–409.

Hellenic Society of Eugenics and Human Genetics. 1981. *Sex Education: Interdisciplinary Symposium: 31 March–1 April 1979, Athens.* Athens: Hellenic Society of Eugenics and Human Genetics.

Hellenic Society for Family Planning, Contraception and Reproductive Health. http://www.oikogeneiakos.gr/index.php?option=com_content&view=article&id=79&Itemid=311&lang=el. Accessed 2 Feb 2014.

Henig, Ruth, and Simon Henig. 2001. *Women and Political Power: Europe Since 1945.* London and New York: Routledge.

Inglehart, Ronald, and Pippa Norris. 2003. *Rising Tide: Gender Equality and Cultural Change Around the World.* Cambridge: Cambridge University Press.

Ioannidi-Kapolou, Elizabeth. 2004. Use of Contraception and Abortion in Greece: A Review. *Reproductive Health Matters* 12 (24): 174–183.

Kattamis, C.N., et al. 1970. Growth of Children with Thalassaemia: Effect of Different Transfusion Regimens. *Archives of Disease in Childhood* 45: 502–505.

Koronaios, G. 1959a. The Agonising Problems Produced By Overpopulation: Is There Enough Space for the Greeks in Greece? A Sensational Discussion Among Seven Top Academics. *Acropolis,* April 26.

Koronaios, G. 1959b. Does Overpopulation Threaten Greece? If the Civilised People Apply Birth Control, the Coloured will Cover the Earth. *Acropolis,* April 29.

Koronaios, G. 1959c. Does Overpopulation Threatens Us? Birth Control Is Not the Number One Problem of Our Country, But Provision of Labour to Everybody. *Acropolis,* May 2.

Kotzamanis, Byron. 2000. *The Demographic Developments During the Interwar Period in the Countries of Eastern Europe and Greece.* Athens: National Center for Social Research.

Kotzamanis, Byron, and Eleftheria Androulaki. 2009. *Elements of Demography.* Volos: University of Thessaly.

Lampadarios, E., and V. Valaoras. 1939. La population grecque vieillit-elle? *Archives Balkaniques de Médecine, Chirurgie et leurs Spécialités* 1: 15–21.

Leon, Sharon M. 2013. *An Image of God: The Catholic Struggle with Eugenics.* Chicago: University of Chicago Press.

Lidz, Ruth W., and Theodore Lidz. 1949. The Family Environment of Schizophrenic Patients. *American Journal of Psychiatry* 106: 332–345.

Löscher, Monika. 2007. Eugenics and Catholicism in Interwar Austria. In *Blood and Homeland, Eugenics and Racial Nationalism in Central and Southeast Europe, 1900–1940*, ed. M. Turda and P. Weindling, 299–316. Budapest: Central European University Press.

Loukopoulos, D. 2011. Haemoglobinopathies in Greece: Prevention Program Over the Past 35 Years. *The Indian Journal of Medical Research* 134 (4): 572–576.

Loukopoulos, D., A. Kaltsoya-Tassiopoulou, and P. Fessas. 1983. Prevention of Thalassemia. *Schweizerische Medizinische Wochenschrift* 113 (40): 1419–1427.

MacKnight, Gerald. 1958. The Great Problem of Overpopulation: An Interview with Bernard Russell. *Kathimerini*, September 3.

Malamos, B., Ph. Phessas, and G. Stamatoyannopoulos. 1962. Types of Thalassaemia-Trait Carriers as Revealed By a Study of Their Incidence in Greece. *British Journal of Haematology* 8 (5): 5–13.

Mauldin, W.P., et al. 1974. A Report on Bucharest. The World Population Conference and the Population Tribune, August 1974. *Studies in Family Planning* 5 (12): 357–395.

McCabe, Lina L., and Edward R.B. McCabe. 2011. Down Syndrome: Coercion and Eugenics. *Genetics in Medicine* 13 (8): 708–710.

Mead, Margaret. 1970. *Culture and Commitment: A Study of the Generation Gap*. New York: Doubleday.

Modell, B., R.H.T. Ward, and D.V.I. Fairweather. 1980. Effect of Introducing Antenatal Diagnosis on Reproductive Behaviour of Families at Risk for Thalassaemia Major. *British Medical Journal* 280: 1347–1350.

Nikolaos Louros Papers and Archive. N. Louros Foundation, Division of History of Medicine, Faculty of Medicine, University of Crete.

Official Government Gazette. 1968. Law 300: About the Premarital Medical Examination Certificate.

Official Government Gazette. 1980. Law 1036: About Family Planning and Other Regulations.

Official Government Gazette. 1982. Law 1250: About the Establishment of Civil Marriage.

Official Government Gazette. 1983. Law 1329: Application of the Constitutional Principle for Sex Equality to the Civil Code, Its Introductory Law, the Commercial Law, the Code of Civil Procedure, and Partial Modernisation of the Civil Code's Ordinances About Family Law.

Official Government Gazette. 1986. Law 1609: Artificial Termination of Pregnancy and Female Health Protection.

Orwell, George. 1949. *Nineteen Eighty-Four*. London: Secker and Warburg.

Pearson, Karl. 1930. *The Life, Letters and Labours of Francis Galton*. London: Cambridge University Press.

Papavassiliou, I.TH. 1954. Intelligence and Family Size. *Population Studies* 7 (3): 222–226.

Rendall, Jane. 1985. *The Origins of Modern Feminism: Women in Britain, France and the United States, 1780–1860.* Basingstoke: Macmillan.

Rosen, Christine. 2004. *Preaching Eugenics: Religious Leaders and the American Eugenics Movement.* Oxford: Oxford University Press.

Singer, M.T., and L.C. Wynne. 1966. Principles for Scoring Communication Defects and Deviances in Parents of Schizophrenics: Rorschach and TAT Scoring Manuals. *Psychiatry* 29 (3): 260–288.

Society of Medical Studies. 1963. Conference of the Hellenic Eugenics Society. *Iatriki* 3 (3): 169–222. Reprint in Nikolaos Louros Papers and Archive.

Spencer, Hamish G., and Diane B. Paul. 1998. The Failure of a Scientific Critique: David Heron, Karl Pearson and Mendelian Eugenics. *The British Journal for the History of Science* 31 (4): 441–452.

Stavropoulos, Alexandre M. 1970. *Bilan analytique et clinique du centre experimental de consultations premaritales et conjugales de la Société Hellenique d' Eugenisme a Athènes.* Louvain: Université Catholique de Louvain.

Stavropoulos, Alexander M. 1977. *The Problem of Reproduction and the Encyclical of the Church of Greece (1937).* Athens.

Stavropoulos, Alexander M. 1981. *The Demographical Problem, Gamily Planning and Abortion: A Theological Approach.* Athens.

Stokes, Shannon C. 1972. Religious Differentials in Reproductive Behaviour: A Replication and Extension. *Sociological Analysis* 33 (1): 26–33.

Sullivan, Harry Stack. 1968. *The Interpersonal Theory of Psychiatry.* New York, London: W. W. Norton.

The Greek Constitution. 1975. *Official Government Gazette*, June 9.

Toth, Adel, Tibor Nyari, and Janos Szabo. 2008. Changing Views on the Goal of Reproductive Genetic Counselling in Hungary. *European Journal of Obstetrics and Gynaecology and Reproductive Biology* 137: 3–9.

Trubeta, Sevasti. 2013. *Physical Anthropology Race and Eugenics in Greece (1880–1970s).* Leiden: Brill.

Turda, Marius. 2009. To End the Degeneration of a Nation: Debates on Eugenic Sterilisation in Inter-War Romania. *Medical History* 53 (1): 77–104.

United Nations. 1974. Report of the United Nations World Population Conference 1974. http://www.un.org/en/development/desa/population/publications/index.shtml. Accessed 14 May 2014.

United Nations. 2002. *Abortion Policies: A Global Review.* New York: United Nation.

Valaoras, Vasilios. 1943. *Elements of Biometry and Statistics. A Demographic Research of the Greek Population.* Athens: Vafiadakis.

Valaoras, Vasilios. 1950. Refined Rates for Infant and Child Mortality. *Population Studies* 4 (3): 253–266.

Valaoras, Vasilios. 1976. *Studies, Titles, Activity and Scientific Works: Additional Text, February 1974–December 1976.* Athens: n.p.

Valaoras, V.G., A. Polychronopoulou, and D. Trichopoulos. 1965. Control of Family Size in Greece: The Results of a Field Survey. *Population Studies* 18 (3): 265–278.

Varika, E. 1987. *The Revolution of the Ladies: The Birth of Feminist Consciousness in Greece 1833–1907.* Athens: Institute of Research and Education of the Commercial Bank of Greece.

Weatherall, D.J. 2010. *Thalassaemia: A Biography.* Oxford: Oxford University Press.

Weatherall, D.J., and J.B. Clegg. 2000. *The Thalassemias Syndromes.* Oxford: Blackwell Scientific Publications.

Wingfield, Nancy M., and Maria Bucur. 2006. *Gender and War in Twentieth-Century Eastern Europe.* Bloomington and Indianapolis: Indiana University Press.

Ziegler, Mary. 2008. Reinventing Eugenics: Reproductive Choice and Law Reform After the World War II. *Cardozo Journal of Law and Gender* 14: 319–347.

CHAPTER 6

Conclusions

The book is a historical account of the eugenics and birth control movement from the 1950s to the 1980s in the Greek and international contexts which is based on original archival research. The originality of the book is attributed to the disclosure of historical information from previously unexplored archives and to the unprecedented combination of Greek, American and British archives. The main argument of the book is to prove the continuity of eugenics during the post-war period which is less discussed by historians. It builds on existing literature on interwar eugenics and post-war birth control by eventually providing a linkage between them, thus expanding our knowledge of reproductive choices and population policies during the post-war period.

The historiographical importance of the book rests on the fact that the case of post-war eugenics, particularly through the prism of the Hellenic Eugenics Society's activities, has not been yet adequately discussed in scholarship. Although there are publications which included elements of its history, this book shed light on the greatly important and yet unexplored archive of Nikolaos Louros, the enduring President of the HES. The revealing information enlightened the published books and articles on post-war Greek eugenics and provided their background thus transforming our understanding of the history of Greek eugenics. Furthermore, the combined study of the archival and published material on issues of eugenics, urbanisation, feminism, family planning, genetic counselling, sex education and many other

© The Author(s) 2019   191
A. Barmpouti, *Post-War Eugenics, Reproductive*
Choices and Population Policies in Greece, 1950s–1980s,
https://doi.org/10.1007/978-3-030-03568-6_6

issues discussed in this book provided the perspective of eugenics in the sociopolitical history of post-war Greece.

The book also added to the current scholarship on the history of post-war eugenics in general. As shown throughout the book and in more detail in Chapters 4 and 5, foreign encouragement for the institutionalisation of eugenics in the 1950s not only was crucial but probably it would have never occurred without support from abroad. It is essential to point out that the history of eugenics continued after the Second World War and was witnessed through the activities of birth control advocates. The existence of post-war eugenics was largely manifested by the correspondence among the Greek, American and British eugenicists which was kept in the archives discussed in this book. Therefore, the book's contribution is equally essential to the history of eugenics, the history of Greek eugenics and the history of birth control.

Corresponding to the title, this book was concerned with the presence of eugenics during the post-war period in the Greek and international contexts, the re-conceptualisation of reproductive choices due to the dissemination of new contraceptive methods and the local and global biopolitics which shaped the population policies during the period from the 1950s to the 1980s.

For the majority of European countries, the given period signified the reconstruction of their identity, society and politics. At the same time, each society tried to understand and adopt the rapid medical and genetic advances which challenged fundamental institutions such as the conjugal relationship and the family. In order to evaluate the sociopolitical events, the historian should try to comprehend and contextualise the multifaceted and groundbreaking transformations that the world experienced after the Second World War. The so-called social progress did not happen by choice but was rather imposed by the race to modernism. In Greece, the modernisation process was often identified with 'westernisation'. In fact, the Greeks made efforts to draw near the Western European countries because they were already three years behind them because of the Civil War that lasted until 1949. Hence, they hurried to abandon their backwardness and assimilate the modern West. Urbanisation was instrumental in this respect.

Although many of the peasants moved to urban centres since the industrial revolutions of the eighteenth and nineteenth centuries, it was during the twentieth century that urbanisation prevailed in the majority of Western countries. Particularly in Greece, the increase of the urbanisation rate in the distribution of population from the 1950s to the 1980s was substantial. By 1980, half of the country's population was gathered in Athens. Undisputedly, the living conditions changed hand in hand with the population movement. The crowded city centres provoked environmental pollution and stressful lifestyles. In addition to the peasants who became workers and employees in the city, the political, scientific and academic elite also gathered in Athens. Thus, major sociopolitical decisions were taken by the parliament located in Athens, social movements firstly occurred there and significant medical advances took place in the big hospitals of the capital. Being the centre of attention, Athens hosted the expression of social tendencies such as eugenics and feminism which were entangled with urbanisation.

The goal of the book was to provide a holistic picture of the time period under consideration by portraying the sociopolitical context in which eugenics resurged and endorsed by the Greeks. This undertaking would not be fulfilled without including the foreign influence which was decisive for the country's modernisation process.

Rather than following the chapter's arrangement, in this chapter we will summarise the main topics of the book in three categories: the unfolding of post-war eugenics, the population policies and demography and the local response to the then novel reproductive choices.

1   POST-WAR EUGENICS

If one should acknowledge a single pattern throughout the historical course of eugenics' presence, this would probably be its malleability. Although claiming that eugenics was a transnational and diachronic phenomenon has become a cliché expression, it is wise to be repeated. Why? Because bearing in mind that its existence was not limited at some point in the past and under certain circumstances, one could identify eugenics ideology in recent debates in contemporary society. It is deemed important to avoid the possibility to relapse to extreme eugenic practices.

Responding to the questions posed in the introduction, in the chapters that followed it we tried to unfold the activities of eugenicists during the less discussed period from the 1950s to the 1980s. This was the time period when globally eugenics embraced population theories and family planning. Eugenics found a space of expression in hot topics, such as overpopulation, environmental changes, exhaustion of resources and immigration. Actually, these were not new concerns in the eugenics agenda; they had been already discussed by Thomas Robert Malthus in late nineteenth century. In reality, the neo-Malthusian eugenics that developed after the Second World War turned its focus on population problems to conceal its underlying racism which was condemned by the international declarations in the aftermath of the Second World War.

Yet, contrary to Malthusian theories, the new eugenics agenda was enveloped by the advance of reproduction technologies, such as in vitro fertilisation and prenatal genetic testing, and was constructed in democratic societies. Paradoxically, however, the proposed eugenic policies for population management expanded to non-Western societies because the problem of overpopulation was transferred to the so-called third world countries. Therefore, we encountered in Chapter 4, wealthy Americans, such as Dr. Gamble, to intervene in counties, such as Pakistan and India, in order to diminish the excessive births. Family planning clinics had been already established in the USA and Britain, and new contraceptive technologies had been tested in countries such as Puerto Rico. The task of American and British eugenicists in the 1950s was to expand the application of their products and expertise in other counties, far away from the USA and Western Europe.

Beyond their alleged concerns about the environmental disasters and limitation of resources that overpopulation entailed, their goal was to eliminate the reproduction of the poor. This was by no means a novelty of post-war eugenics; it was one of the essential components of eugenics rhetoric since the Galton era. Similar to the British eugenicists' efforts against pauperism in the beginning of the twentieth century, during the post-war period we encountered the collective effort against the proliferation of the poor in overpopulated countries. For instance, Gamble's associates repeatedly visited poor countries in Africa, Latin America, the Middle East, India and Pakistan. Their work included the foundation of birth control clinics and dissemination of eugenics propaganda. Obviously, their goal was not to help families to arrange the number of births and the space among them but to limit the reproduction of poor, lower classes. To this end, they employed environmental disaster and the spread of diseases to justify their involvement in the population management of the 'third world' countries.

Although Greece was not included in these countries, Chapters 3 and 4 brought to light the hidden history of foreign influence in the birth of the Greek post-war institutionalisation of eugenics. Foreign involvement was manifold and materialised by the participation of American and British eugenicists. According to our research, the American demographer Pascal K. Whelpton was the first to contact Greek physicians in 1952 and discuss about population management and eugenics. His speech in Athens triggered the formulation of the first group of supporters of eugenics at the premises of the Athens Medical Association. Whelpton's success in influencing the Greek physicians to form a society of eugenicists was facilitated by the sociopolitical circumstances of the country and the low health level of its citizens. Whelpton inspired the Greek physicians by offering them a solution to the demographical problems of Greece. As suggested in many parts of this book, the tendency to think of themselves as safeguards of the social and health prosperity of the country was a common feature among eugenicists in different localities and time periods. Therefore, the Greek eugenicists proceeded with the formal establishment of the Hellenic Eugenics Society inspired by Whelpton's urge for proper population management according to the principles of eugenics.

In the process, we witnessed the development of a network among Greek, American and British eugenicists who retained their close relationship by correspondence and mutual visits. According to the archives, there was frequent correspondence among Athens, New York and London in particular during the 1950s and early 1960s. The relationship was not limited to correspondence but escalated to personal visits both in Greece and in other European countries and the development of friendships. A visit in the USA was not verified by the archives, but it would be possible for Kanavarioti to have visited birth control clinics there after her movement in the USA in late 1950s. Highlighting the most important collaborations among this network, these were: Whelpton's visit in 1952; Kanavarioti's election as a member of the Governing Body of the IPPF in September 1954; Gamble's first shipment of contraceptives to Louros in February 1955 and Houghton's overall assistance to Kanavarioti but particularly during the preparation of Louros' inaugural speech on eugenics in March 1955. Whelpton's visit was of utmost important because it was the first recorded visit by an American population expert and also because it aided the institutionalisation of eugenics in Greece. Kanavarioti's election as member of the Governing Body of the IPPF was equally important because it confirmed the collaboration between the HES and the IPPF and verified the assumption that Kanavarioti was the most active person in pursuing the international recognition of the HES. Furthermore, her participation in the historical, first world population conference in Rome should not be overlooked. Kanavarioti's work was highly appreciated by Vera Houghton. She assisted her with every possible way, in particular during the preparation of Louros' first public lecture in March 1955. As was discussed, the success of this lecture impacted both in the local and global contexts. Last but not least, the first shipment of contraceptives from Gamble to Louros was a hallmark in the history of contraception in Greece. It was the time that contraception passed the threshold of theory to practice.

What is most important is the fact that none of these would came to light without the study of a number of archives; it is the invaluable worth of the archives and their comparative study that added value to this book. It became thus possible to answer the question: What was the role of eugenicists during the period from the 1950s to the 1980s? First and foremost, this book aimed at proving that eugenics continued to exert influence in many social contexts even after its condemnation in late 1940s. We argued that post-war eugenicists continued to form national societies and develop international bonds similarly to the beginning of the twentieth century albeit more cautiously and under the pretence of saving the planet from an imminent catastrophe caused by overpopulation. What is more, the means to control heredity transformed from coercive sterilisations and selective marriages to the more scientific solutions of the use of female contraceptives and prenatal genetic diagnosis. Not only was the Nazi legacy that made eugenicists to abandon the former practices of direct state intervention to the individual reproductive choices but also the social transformations and the biomedical advances that occurred during this period.

In post-war Greece, eugenics was interpreted as a way to reconstruct the society as a whole by permitting the couples and women to plan their families. Yet, this was to be accomplished according to their standards. They aimed at reaching both the scientific circles and the wide public in order to influence them by persuading them to adopt a eugenics mentality. Therefore, although the HES was the primal eugenics establishment and labelled as such, it was not the only place to find proponents of eugenics. As was discussed in Chapter 3, scientific groups such as the NUSE also promoted eugenics and hygiene. In addition, many of its members were also affiliated with the HES. Moreover, in Chapter 4, Gamble and his team corresponded and visited associations and gynaecologists outside the HES. Among them, the relationships with women's clubs and the PIKPA dominated. Again, it was Kanavarioti who introduced the American field workers to people working in these institutions. For instance, Kanavarioti introduced Gates to physicians of the PIKPA and members of the National Council of Greek Women. Moreover, Gamble's keen interest in diminishing the number of poor people motivated him to collaborate with Dr. Tsacona who worked in Thessaloniki, the second most populated city in Greece. It seems that he wanted to exploit every possible contact to achieve his target.

The study of the correspondence between Gamble and Tsacona shed light on the one hand to the difficulty of passing contraceptives through the Greek authorities and on the other hand Gamble's persistence and determination to find a way to achieve his target. In this context, he repeated many times that the supplies were destined to the poor people of Thessaloniki. Moreover, the collaboration with Tsacona would offer him more results in the effectiveness of contraceptives. Gamble requested feedback from all the countries where his contraceptives were distributed. He and his associates collaborated with companies that produced contraceptives and recommended their products. This became particularly evident when Gates communicated with Greek eugenicists but also with the representative of the Greek company 'Nicolakis' who wanted to import contraceptives.

On the one hand, many Greek birth control advocates positively corresponded to foreign and local geopolitics based on eugenic rationale. On the other hand, however, state authorities impeded their activities. First of all, there were laws prohibiting the import and distribution of contraceptives; secondly, it was also prohibited to open birth control clinics; thirdly, politicians were afraid to pass any form of legislation which potentially resulted in birth limitation. It was instilled in the politicians' minds that the decrease of births would result in fewer future soldiers and such a possibility should be avoided. Although this argument was turned down by the Colonel Merenditis in one of the HES conferences, as was discussed in Chapter 5, it was not abandoned until the 1980s when the new legislation regarding contraception and abortion was implemented.

Coupled with a falling birth rate, the problematic demography of Greece eventually exerted more influence to eugenicists than their foreign counterparts. Therefore, the distribution of contraceptives was never realised as Gamble had envisioned. Added to this, the IPPF found no official place in Greece before the legalisation of female contraception and the state's permission of establishing family planning clinics in 1980s.

2   POPULATION POLICIES

Demography had been constantly debated in Greece since the beginning of the twentieth century, as became evident more or less in all chapters. In the second chapter, the various changes in demography and population policies primarily due to wars and epidemics were illustrated. While the First and Second World Wars resulted in population loss, the Balkan Wars and the catastrophe of Asia Minor resulted in population growth. In this context, it should be highlighted that population loss occurred in a slow pace while the addition of population masses took place abruptly and violently. As a result, the addition of a large quantity of population and territory after the Balkan Wars and the influx of refugees in the aftermath of the defeat in Asia Minor were not unproblematic. In fact, in all cases the state found itself unable to correspond to the circumstances quickly and effectively because the relocations of the population took place in a country where political polarity and instability prevailed for decades. By the 1930s, the refugees from Asia Minor had been integrated into the total population. With the exception of the annexation of Dodecanese islands in 1948, Greece did not experienced any other massive growth or loss in population. The immigration waves in Western Europe, USA, Canada and Australia during the 1950s–1960s were significant but not as crucial as the population movements caused by the wars.

The unpredictability of population changes however caused the insurmountable obstacle of the fear of governments to pass any legislation that might provoke birth limitation. This was the starting point of pro-natalist policies together with the high rise of induced abortions. Not least, the geographical position of Greece added anxiety to the governments. As sufficiently discussed throughout the book, the fear of a possible military invasion from a neighbouring country was the reason why the state authorities refused to legalise female contraceptives until as late as the eighties. The efforts of Greek eugenicists to convince the authorities to repel the legislation which prohibited the distribution of contraceptives were unfolded in Chapter 4. Even Louros who was such an influential personality faced the objection of the Ministry of Health in the 1950s. Much later, in 1968, he achieved the establishment of the experimental institution of Premarital Advisory Centre at Alexandra Maternity Hospital. However, the victory of eugenicists was sealed by Spyros Doxiadis, another President of the Hellenic Eugenics Society, who signed the law 1036 in 1980 permitting the foundation of Family Planning Centres in public hospitals throughout the country.

The most alarming population problem in Greece was induced abortions. Either legally after 1986 or illegally earlier, the rate of abortions in Greece was very high. This was not hard to explain. Before the free use of contraceptives, women's choices for controlling their fertility were narrowed to abortion or abstinence. Therefore, to a certain extent, the high rate of abortions was reasonable. Couples often chose abortion to avoid an unwanted pregnancy or to be able to control the size of their family.

After the legalisation of contraceptives, there were two major obstacles in their distribution, as was argued in Chapters 4 and 5. Firstly, many gynaecologists discouraged women to use them because they would lose their profit from performing abortions. Although opposite to certain standards of medical deontology, this was an undeniable fact in Greece. Secondly, an equally crucial factor contributing to the large number of induced abortions was the absence of sex education. Chapter 5 included the public conference on sex education which was organised by the HES in which the complexity of the issue was portrayed by the participation of many experts in the field. Even in the present day, sex education is not included in school curricula in Greek public schools. Unless the policy-makers pass a law supporting the teaching of the course, sex education will rest at the hands of parents and peers and abortion rates will continue to be high.

From the standpoint of the Orthodox Church, abortion was equated with homicide. Man was not entitled to rule over life and death, and murder was the ultimate sin. In addition, according to the Orthodox teaching, human life begins at conception not at birth. Bearing the above in mind, abortion was strictly condemned by the Church. The fact that Orthodox Christianity was the predominant religion in Greece added complexity to the issue of abortion because all these couples and gynaecologists who performed numerous abortions presumably were believers who opposed such a fundamental teaching. At least, the Church organised sex education lectures to inform its believers whereas the state authorities did not.

Therefore, induced abortions continue to be a thorny issue in Greek demography. Fortunately, the law which legalised abortion under certain circumstances permitted women to perform it in a public hospital. This way the operation became safe and legal. To a great extent, Greek eugenicists argued that the abortion rates would have dropped if contraception was allowed. In retrospect though, female contraception was not largely embraced by Greek women. The condom remained the prevailing contraceptive method, and the problem of induced abortions continued to concern demographers.

3   REPRODUCTIVE CHOICES

As was highlighted throughout the book and particularly in the fifth chapter, during the decades from the 1950s to the 1980s the reproductive choices underwent several changes. From the prohibition of female contraceptives to the legalisation of abortion, the control of the female body was entangled with both eugenics and feminism. Eugenicists and feminists, bearing their own ideology and principles, influenced national and international politics in shaping legislation regarding reproduction. In this book has been analysed the mutual interaction between the establishment of eugenics and the modernisation of the Greek state's legislation regarding human reproduction, marriage and demography. Bearing witness of the involvement of eugenicists in political positions and the contribution of politicians in eugenic conferences, it becomes clearly evident that on the one hand, eugenicists were not an isolated group of academics but socially respected individuals, and on the other hand, that politicians sought ways and means to modernise their legislation according to social currents.

Therefore, the legalisation of female contraceptives and abortion came with no surprise in the late 1970s. Undoubtedly, the participation of charismatic women in the sociopolitical reconstruction of the country was of utmost importance and should not be overlooked. Either medical professionals or political activists, such as Maro Kanavarioti and Lina Tsaldaris, they played an important role in transforming the reproductive choices of the Greek women from the illegally induced abortion as the only choice of family planning to the wide spectrum of contraception. However, it should be highlighted that the motives behind the support of birth control practices were very different for women and feminists and for eugenicists. The former supported birth control because they pursued female emancipation, whereas the latter because they wanted to control the quality and quantity of the overall population. Moreover, as was discussed in Chapter 4, foreign eugenicists aided their Greek counterparts aiming at the control of heredity and the elimination of the poor population of Greece, whereas the Greek eugenicists aimed at the regeneration of the society and the elimination of illness regardless of social class.

The process of receiving and distributing female contraceptives, even though at a minor level, as happened by Louros and other eugenicists, was the first attempt to familiarise the Greek women with this kind of reproductive choices. Due to the fact that female contraceptives were illegal until 1980s, relevant publications before that time do not exist. Therefore, this book offered this unique information which was found in the unexplored Louros' and Gamble's archives. It was important to know that female contraceptives were not firstly imported in Greece in the late 1970s but much earlier, in the 1950s. Moreover, it was equally important that contraceptives were shipped by Gamble and received by Greek eugenicists. In addition, personal correspondence revealed that not only was the HES interested in female contraception but also women's clubs and medical professionals who worked closely with women, such as people working at the PIKPA. There is not concrete evidence however that they received and distributed them by Gamble's associates. Although they might never distributed them, the interest on the side of medical professionals and on the side of their patients was certainly verified. Above all, women desired to control their reproduction and claimed their right to control their body, their life and their family.

In conclusion, the approximately thirty-year period under examination in this book was a transitional period regarding eugenics, reproductive choices and population policies both in the international and local levels. Eugenicists tried to survive the post-Nazi era by promoting family planning. However, this was not always possible or as effective as anticipated. At the same time, social fermentations mainly caused by the rise of feminism and urbanisation were important factors in shaping anew the reproductive choices. Following the streamline of sociopolitical changes, at least in the European context, it was observed a concurrent legalisation of abortion and groundbreaking changes in the legislation about marriage and parenting. The same period also hosted a number of important biomedical advances preparing the field for present-day's reprogenetics. For the reasons stated above, this time period was very interesting and worth further study. Although for some, eugenics is an extinct ideology, this book hopes to have inspired researchers to dig deep in archives and publications of the time and bring to light more information on the involvement of international organisations of eugenicists to national contexts and eventually draw the global picture of post-war eugenics. As widely discussed and justified, eugenics did not disappear and actually never stopped to gain supporters across geographies and societies. In fact, we could safely argue that at present is more achievable than any past period due to the available biotechnology to affect the selection of births and human breeding.

In light of reproduction genetics, the modern pregnant woman rarely chooses to leave things to chance. On the contrary, she might find herself guilty of not choosing the best for her offspring. Of course, it was a blessing to be able to prevent one's family and the world from genetic diseases, but the lines between prevention and eugenics were often blurred. It was extremely difficult to set guidelines on what to choose and what to avoid with regard to the physical and mental state of the embryo. The advance in reproduction genetics offered man the advantage of prenatal knowledge. At the same time however the society indirectly prohibited their freedom of choice. For instance, the gynaecologist and the society most probably would not agree with a couple's decision to bring to life a child with Down's syndrome. So, would it be accurate to argue that today the reproductive rights are respected whereas in the past they were not? Even though state-imposed eugenics was limited after the Second World War, socially imposed eugenics succeeded it.

Last but not least, human enhancement technologies continuously develop and draw closer to the creation of a 'superhuman' with unprecedented mental and physical abilities. The creation of the post-human that transhumanism envisions, either achievable in the near future or not, still adheres to the fundamental eugenic ideals of the control of heredity, human enhancement and social perfection.

INDEX

A

Alexandra Maternity Hospital, 59, 62, 64, 66, 67, 92, 103, 107, 111, 112, 117, 118, 121, 122, 124, 127, 166, 171, 199

American Eugenics Society (AES), 4, 100, 101

Anaemia, Mediterranean, 171–173, 177, 181–184

Athens Medical Association, 36

B

Blacker, Carlos Paton, 68

British Eugenics Society (BES), 4, 12, 56, 67, 71, 85–87, 94, 99, 102

Brush, Dorothy, 12, 65, 70, 86, 87, 95, 96, 100, 103

C

Church, Orthodox, 51, 52, 72, 100, 113, 123, 135, 161, 166, 167, 180, 199

Counselling, 173
  genetic counselling, 176
  medical counselling, 175

D

Doxiadis, Spyros, 7, 34, 52, 54, 55, 59, 65, 75, 78, 79, 85, 96, 98, 117, 150, 152, 174, 199

F

Family law, The, 9, 158, 160, 161, 164

Feminism, 17, 160, 191, 193, 200, 201

G

Galton, Francis, 34

Genetics, 2, 4, 5, 9, 10, 13, 14, 17, 38, 55, 56, 58, 112, 113, 152, 163, 170, 202

H

Hereditary diseases, 17, 58, 142, 169, 170, 177, 181, 182

Hygiene, 1, 6, 7, 11, 15, 22–28, 30–33, 38–40, 45–47, 62, 72–74, 76, 78, 90, 106, 116, 117, 120, 123, 124, 134, 142, 153, 155, 165, 166, 173, 196

M

Malthus, Thomas Robert, 142

  Malthusian, 142

  neo-Malthusianism, 87

Medical Council, 24

N

Neo-Malthusianism, 8, 119

P

Propaganda, 13, 39, 45, 51, 58, 61, 62, 64, 74, 102, 104, 105, 111, 119, 121, 126, 127, 194

S

Sanger, Margaret, 86

Sex education, 8, 46, 94, 113, 161, 165–168, 191, 199

Sterilisation, 5, 69, 74, 75, 90, 100, 105, 117, 124, 136, 177, 181, 196

Stone, Abraham, 86, 87, 98, 102

Student health card, 15, 32

V

van Vleck, Joseph, 61, 63–65, 86, 97, 98, 103, 106, 125

Vogt, William, 50

W

Whelpton, Pascal, 48

World Population Conference, Rome, 96, 97, 102, 152, 153, 196

The manufacturer's authorised representative in the EU is Springer
Nature Customer Service Centre GmbH, Europaplatz 3, 69115 Heidelberg,
Germany. If you have any concerns regarding our products, please
contact ProductSafety@springernature.com

Printed and bound by CPI Group (UK) Ltd, Croydon, CR0 4YY
23/04/2026
02095601-0001